S.S.F. PUBLIC LIBRARY
WEST ORANGE

Y0-AGT-787

621.38
L

243425   Copy

Author: Lenk, John D
Title: Handbook of basic electronic
         troubleshooting

South San Francisco Public Library

10   RULES

A fine of 2 cents a day will be charged on books
kept overdue.
No books will be issued to persons in arrears for
fines.
Careful usage of books is expected, and any soiling,
injury or loss is to be paid for by the borrower.

Keep This Card in The Pocket

APR   1981

Handbook of Basic Electronic Troubleshooting

Handbook
of Basic
Electronic
Troubleshooting

JOHN D. LENK

Consulting Technical Writer

PRENTICE-HALL, INC., *Englewood Cliffs, New Jersey*

Library of Congress Cataloging in Publication Data

Lenk, John D date
  Handbook of basic electronic troubleshooting.

  1. Electronic apparatus and appliances—Maintenance
and repair. 2. Electronic apparatus and appliances—
Testing. I. Title.
  TK7870.L457   621.381'028   76-7538
  ISBN 0-13-372482-4

© 1977 by Prentice-Hall, Inc.
Englewood Cliffs, New Jersey

All rights reserved. No part of this book
may be reproduced in any form or by any means
without permission in writing from the publisher.

10 9 8 7 6 5 4 3 2 1

Printed in the United States of America

Prentice-Hall International, Inc., *London*
Prentice-Hall of Australia, Pty. Ltd., *Sydney*
Prentice-Hall of Canada, Ltd., *Toronto*
Prentice-Hall of India Private Limited, *New Delhi*
Prentice-Hall of Japan, Inc., *Tokyo*
Prentice-Hall of Southeast Asia Pte. Ltd., *Singapore*

To my wife *Irene* from *El Autor*

243425

E.S.F. PUBLIC LIBRARY
WEST ORANGE

243425

C.B. PUBLIC LIBRARY
WEST ORANGE

Contents

Index   **233**

Preface

As its title implies, the *Handbook of Basic Electronic Troubleshooting* is devoted exclusively to *basic* troubleshooting procedures. That is, this handbook describes procedures for analyzing trouble symptoms, localizing faulty circuits, and isolating defective components in basic types of electronic equipment and circuits.

This handbook is an ideal companion to the author's highly successful *Handbook of Practical Sold-State Troubleshooting*. However, no direct reference to the solid-state book is required to make full use of this handbook.

The solid-state book is written on the assumption that you, the reader, are thoroughly familiar with basic electronic test equipment. For *Handbook of Basic Electronic Troubleshooting*, no such assumptions are made. Instead, this handbook describes what test equipment is available, how it works, and how it is used in troubleshooting.

A knowledge of test equipment and procedures is vital for really efficient troubleshooting. For example, one of the early steps in most troubleshooting procedures is to monitor operation of the equipment or circuit with various pieces of test equipment. If you do not understand operation of the test equipment, or the *connections* required to monitor circuit, you will be hopelessly lost.

We do assume in this handbook that you are familar with basic electronics but have no particular knowledge of test equipment and no practical experience in troubleshooting. Thus, the handbook serves as an ideal first book of troubleshooting for the beginner. However, the *basic techniques* described here apply to all types of electronics. As such, the handbook can be useful for advanced technicians and engineers.

One major advantage of this handbook is that it teaches you to trouble-shoot electronic equipment with *very limited* service literature. One of the drawbacks with other troubleshooting books is that they assume that full military-type service literature will be provided for all equipment. This is almost never the case in the real world of electronic service work. Instead, the technician must get along with the most fragmentary service literature and often with no literature whatsoever. This handbook provides you with techniques to overcome such problems.

The book presents many *troubleshooting examples* in *step-by-step form*. Steps that *could* be taken and *should* be taken are described. In many troubleshooting situations, the could-be steps are almost as logical as the should-be steps. This book describes the difference between could-be and should-be approaches, which generally makes the difference between the "textbook" technician and the really top-drawer troubleshooter.

The handbook is divided into three chapters. Chapter 1 provides an introduction to troubleshooting. Such subjects as basic troubleshooting sequence, failure-symptom analysis, and localizing troubles to modules, circuits, and specific components are covered.

Chapter 2 is devoted entirely to test equipment that is used in electronic troubleshooting. No single chapter could list all items of troubleshooting test equipment, much less describe such a piece of equipment in full detail. However, Chapter 2 describes both the purpose and the operating principles related to all types of troubleshooting test equipment. By concentrating on generalized circuits, at the level of the block diagram or simplified schematic diagram, the specific circuit of any basic troubleshooting device can be understood.

Chapter 3 deals with basic circuit (and equipment) troubleshooting. This chapter describes how the basic troubleshooting techniques discussed in Chapter 1 are combined with the practical use of test equipment described in Chapter 2 to locate specific faults in various types of electronic circuits.

Because alignment and adjustment, as well as testing, are part of trouble-shooting, Chapter 3 describes all basic adjustment procedures for the circuits and equipment discussed. This is followed by specific examples of trouble-shooting for the circuits or equipment. Thus, it is not necessary to refer to other publications for basic test and adjustment data.

The author has received much help from many organizations and individuals prominent in the field of electronics. He wishes to thank them all.

The author also wishes to express his appreciation to Mr. Joseph A. Labok of Los Angeles Valley College for his help and encouragement.

JOHN D. LENK

Handbook of Basic Electronic Troubleshooting

1

Introduction
to Troubleshooting

Troubleshooting can be considered a step-by-step logical approach to locating and correcting any fault in the operation of a piece of equipment. In the case of electronic troubleshooting, there are seven basic functions required.

First, you, the technician, must study the equipment, using service literature, schematic diagrams, and so forth, to find out how each circuit works when operating *normally*. In this way, you will know in detail how a given piece of electronic equipment or system *should* work. If you do not take the time to learn what is normal, you will never be able to distinguish what is abnormal.

Second, you must know the function of all equipment controls and adjustments and how to manipulate them. It is difficult, if not impossible, to check out a piece of equipment without knowing how to set its controls. Also, as equipment ages, readjustment and realignment of critical circuits are often required.

Third, you must know how to interpret service literature and how to use test equipment. Along with good test equipment that you know how to use, well-written service literature is your best friend.

Fourth, you must be able to apply a systematic, logical procedure in order to locate the trouble. Of course, a procedure that is logical for one type of equipment is quite illogical for another. For that reason, we shall discuss logical troubleshooting procedures for various types of equipment, as well as basic troubleshooting procedures.

Fifth, you must be able to analyze logically the information provided by an improperly operating piece of equipment or system. The information to be analyzed may be the equipment's performance (e.g., the appearance of the picture on a television screen) or indications taken from test equipment (e.g., voltage and resistance readings). Either way, it is the analysis of the information that makes for logical, efficient troubleshooting.

Sixth, you must be able to perform complete checkout procedures on the repaired equipment. Such a checkout may require only simple operation (e.g., switching through all channels on a television set and checking the picture). At the other extreme, the checkout may involve complete realignment of the television set. Either way, a checkout is always required after troubleshooting. One reason for the checkout is that there may be more than one trouble. For example, an aging part may cause high current to flow through a resistor, resulting in the burnout of the resistor. Logical troubleshooting may lead you quickly to the burned-out resistor. Replacement of the resistor will restore operation. However, only a thorough checkout will reveal the original high-current condition that caused the burnout. Another reason for after-service checkout is that the repair may have produced a condition that requires readjustment. A classic example of this occurs when replacement of a part changes circuit characteristics; for example, a new transistor in an intermediate-frequency (IF) stage may require complete realignment of the IF stage.

Seventh, you must be able to use the proper tools to repair the trouble. As a minimum for electronic repair, you must be able to use soldering tools, wire cutters, long-nose pliers, screwdrivers, and socket wrenches. If you are still at the stage where any of these tools seem unfamiliar, you are not ready for basic troubleshooting.

In summary, before starting any troubleshooting job, ask yourself these questions: Have I studied all available service literature to find out how the equipment works? Can I operate the equipment properly? Do I really understand the service literature, and can I use all the required test equipment properly? Using the service literature and/or previous experience on similar equipment, can I plan a logical troubleshooting procedure? Can I analyze the results of operating checks logically, as well as check out output procedures involving test equipment? Using the service literature and/or experience, can I perform complete checkout procedures on the equipment, including realignment, adjustment, and so forth, if necessary? Once I have found the trouble, can I use common hand tools to make the repairs? If the answer to any of these questions is no, you are simply not ready to start troubleshooting. Start studying instead!

1-1. OVERALL TROUBLESHOOTING PROCEDURE

The troubleshooting functions discussed thus far can be divided into four major steps:

1. *Determine* the trouble symptoms.

2. *Localize* the trouble to a functional unit.

3. *Isolate* the trouble to a circuit.

4. *Locate* the specific trouble, probably to a specific part.

The remaining sections of this chapter are devoted to these four steps. Before going into the details of the steps, let us examine what is accomplished by each.

1-1.1 Determining Trouble Symptoms

Determining symptoms means that you must know what the equipment is supposed to do when it is operating normally and, in addition, that you must be able to recognize when that normal job is not being done. Most electronic equipment or systems have operational controls, indicating instruments, or other built-in aids for evaluating their performance. Even the simple transistor radio has a loudspeaker, an ON-OFF switch, a station selector, and a volume control. You must analyze the normal and abnormal symptoms produced by the equipment's built-in indicators in order to formulate the following questions: What is this equipment supposed to do? How well is this job being done? Where in the equipment could there be trouble that will produce these symptoms?

The determining-symptoms step does not mean that you should charge into the equipment with screwdriver and soldering tool, nor does it mean that test equipment should be used extensively. Instead, this first step in troubleshooting deals mainly with using your powers of observation (including a visual inspection of the equipment and, for certain types of complex equipment, the use of test equipment to observe input and output wave forms, measure the power output, and so forth) and your knowledge of what the equipment is supposed to do and how it works. In general, for the less complex equipment, the determining-symptoms step involves noting both the normal and the abnormal *performance indications*, manipulating the equipment's operational controls to gain further information, and correlating the symptoms.

At the end of the determining-symptoms step, you definitely know that something is wrong and have a fair idea of what the trouble is, but you do not know just what area of the equipment is faulty. This is established in the next step of troubleshooting.

1-1.2 Localizing Trouble to a Functioning Unit

Most electronic equipment can be subdivided into units or a eas that have a definite purpose or function. The term *function* is used here to denote an electronic operation performed in a specific area of a piece of equipment. For example, even a simple radio receiver has a radio-frequency (RF) section, an IF section, and an audio section. In the case of a more complex piece of equipment such as a television set, the functions can be subdivided into audio, video, tuner, picture tube, and power supply. These functions, when combined, make up an equipment *set* and cause the set to perform the electronic purpose for which it was designed.

In order to localize the trouble systematically and logically, you must know the functional units of the equipment and correlate all the symptoms previously determined with those units. Thus, the first consideration in localizing the trouble to a functional unit really is a valid estimate of the area in which the trouble *might* be in order to cause the indicated symptoms. Initially, several technically accurate possibilities may be selected as the probable trouble area.

Electronic troubleshooting involves the extensive use of diagrams. These include *functional block diagrams, overall schematics,* and *wiring diagrams,* as shown in Figs. 1-1 to 1-3, respectively.

The block diagram (Fig. 1-1) shows the *functional relationship* of *all major sections* in a complete piece of equipment. The block diagram is thus the most logical source of information to use when localizing trouble to a functioning unit or section. Unfortunately, not all service literature includes a block diagram. Therefore, it may be necessary to use the overall schematic.

The schematic (Fig. 1-2) shows the *functional relationship* of *all parts* of a complete piece of equipment. Such parts include all resistors, capacitors, transistors, diodes, and so forth. Generally, the schematic presents

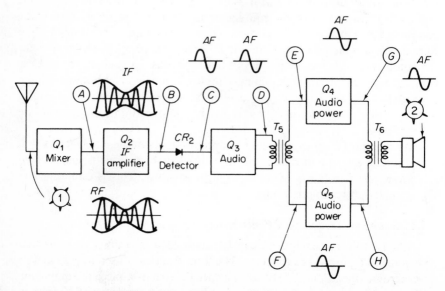

FIGURE 1-1   Functional block diagram of receiver

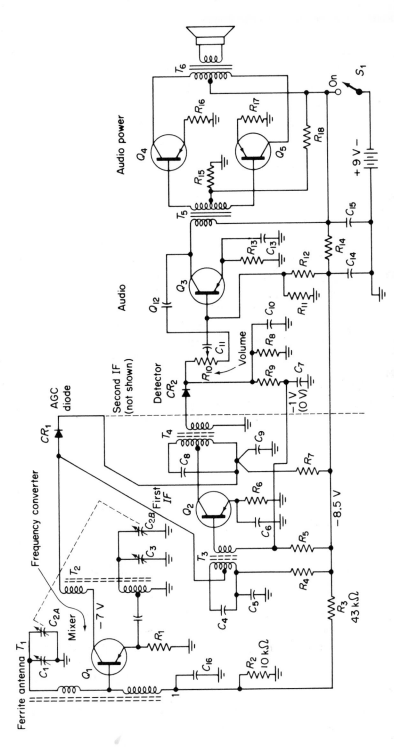

FIGURE 1-2 Schematic diagram of receiver

5

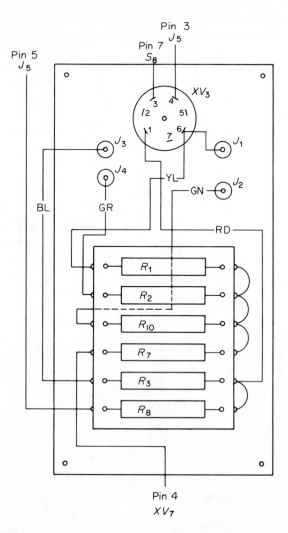

FIGURE 1-3 Partial wiring diagram showing physical relationship of parts and wiring

too much information (much of it not directly related to the specific symptoms you have noted) to be of maximum value during the localizing step. If you tried to take all the details into consideration, it might become extremely difficult for you to make decisions regarding the probable trouble area. However, the schematic is very useful in later stages of the total troubleshooting effort and when a block diagram is not available.

In comparing the block diagram and the schematic during the localizing step, note that each transistor shown on the schematic (Fig. 1-2) is represented as a block on the block diagram (Fig. 1-1). This relationship is typical of most service literature.

The wiring diagram (Fig. 1-3) shows the *physical relationship* of *all parts*, as well as point-to-point wiring. For this reason, the wiring diagram is the least useful source of information when you are localizing trouble. However, the wiring diagram is most useful when you are locating *specific* parts during the repair phase of troubleshooting.

To sum up, it is logical to use a functional block diagram instead of a schematic or wiring diagram when you want to make a valid estimate of the probable trouble areas (the faulty functioning units). The use of a block diagram also permits you to use a troubleshooting technique known as *bracketing*. If the block diagram includes *major test points*, as it should in well-written service literature, it will also permit you to use test equipment as an aid in narrowing the choices of probable trouble areas to a single functional unit.

1-1.3 Isolating Trouble to a Circuit

After the trouble is localized to a single functional unit, the next step is to isolate the trouble to a circuit in the faulty unit. To do this, you consider the signal paths in the circuitry that contains the indicating instruments or other built-in aids that point to *abnormal performance*. By concentrating on this circuitry and ignoring the circuits that produce normal indications, you narrow down or isolate the limits of the possible trouble. For example, assume that you are troubleshooting a communications set that contains a transmitter, a receiver, and a common power supply. If the receiver is operating normally but the transmitter is not, you can ignore both the receiver circuits and the common power supply circuits and concentrate on the transmitter circuits.

The isolating step in troubleshooting involves the use of test equipment such as meters, oscilloscopes, and signal generators for *signal tracing* and *signal substitution* in the suspected faulty area. By making valid educated estimates and using applicable diagrams, bracketing techniques, signal tracing, and signal substitution properly, you can systematically and logically isolate the trouble to a single defective unit.

Repair techniques or tools to make necessary repairs are not used until after the specific trouble is located and verified. That is, you still do not charge into the equipment with soldering tool and screwdriver. Instead, you now try to isolate the trouble to a specific defective circuit so that it can be repaired.

Observations and decisions are made at this time, but operational evaluations are not. More specifically, the observations are now based on the indications of external test equipment used for signal tracing or signal substitution, and the decisions relate to whether these indications are normal or abnormal, based on your knowledge of how the equipment works.

1-1.4 Locating the Specific Trouble

Although this troubleshooting step refers only to locating the specific trouble, it includes a final analysis or review of the complete procedure and the use of repair techniques to remedy the trouble once it has been located. This final analysis will enable you to determine whether some other malfunction caused the component to become faulty or whether the component located is the actual cause of the equipment trouble.

When you are trying to locate the trouble, inspection using the senses of sight, smell, hearing, and touch is very important. This inspection is usually performed first in order to gather information that may more quickly lead to the defective component. Among the things to look for during the inspection using the senses are burned, charred, or overheated components; arcing in the high-voltage circuits; and burned-out parts.

In equipment where it is relatively easy to gain access to the circuitry, a rapid visual inspection should be performed first. Then, the active device (vacuum tube or transistor) can be removed from the circuit and checked. A visual inspection is always recommended as the first step for all solid-state equipment and for most vacuum-tube equipment. A possible exception would be equipment in which access to circuit parts is difficult but vacuum tubes can be easily removed and tested (or substituted).

The next step in locating the specific trouble is the use of an oscilloscope to *observe wave forms* and a meter to *measure voltages*. Finally, in most cases, a meter is used to make *resistance* and *continuity checks* in order to pinpoint the defective component.

After the trouble is located, a final analysis of the complete troubleshooting procedure should be made to verify the trouble. The trouble should then be repaired, and the equipment checked out for proper operation.

Note that in the service literature for home entertainment equipment such as a radio and a television set, the wave forms, voltages, and resistances are given on the schematic. For military-type equipment, such information is provided in chart form along with the service literature. Either way, you must be able to use test equipment to observe the wave forms and make the measurements. For this reason, the function and use of test equipment during troubleshooting is discussed frequently throughout this book.

1-1.5 Developing a Systematic, Logical Troubleshooting Procedure

The development of a systematic and logical troubleshooting procedure requires:

a step-by-step approach to the problem

knowledge of the equipment

interpretation of test information

use of information gained in each step

Some technicians feel that knowledge of the equipment involves remembering past equipment failures, as well as remembering data such as the location of all test points and all adjustment procedures. This approach may be valid if you troubleshoot only one type of equipment, but is has little value in the development of a basic troubleshooting procedure.

It is true that recalling past equipment failures may be helpful, but you should not rely upon the possibility that the same trouble will be the cause of a given symptom in every case. In electronic equipment, there are many possible troubles that can give approximately the same symptom indications.

Similarly, you should never rely *entirely* upon your memory of adjustment procedures, test-point locations, and the like in approaching any troubleshooting problem. This is one of the reasons for having service literature that contains diagrams and information about the equipment. The use of service literature and the specific types of information presented in it will be discussed more fully later in this book. The important point for you to learn now is to be a systematic, logical troubleshooter, not a memory expert.

1-2. RELATIONSHIP AMONG TROUBLESHOOTING STEPS

Thus far, we have established the overall troubleshooting procedure. Now, let us make sure that you understand how the troubleshooting steps fit together by analyzing their relationships.

Figure 1-4 is a block diagram showing the relationships among the steps; it also shows specific things you must do or must use to complete the requirements of each step satisfactorily.

The first step, determining the symptoms, requires use of the senses, observation of equipment performance, previous knowledge of equipment operation, manipulation of the operational controls, and recording of notes. Determining the symptoms presupposes the ability to recognize improper indications, to operate the controls properly, and to record the effect that the controls have on trouble symptoms.

The second step, localizing the trouble to a functional unit, depends upon the information observed and recorded in the first step, the use of a functional block diagram (or possibly the schematic), and reasoning. During the second step, ask yourself the question: What functional area could cause the indicated symptoms? Then, bracketing (narrowing down the probable deductions to a single functioning unit) is used along with the test equipment to pinpoint the faulty function. The observations in this step

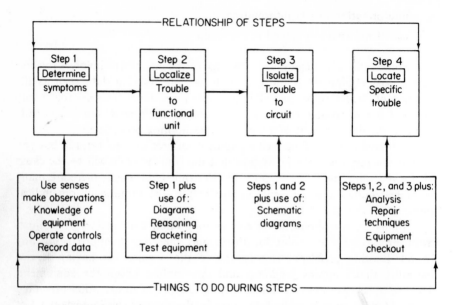

RELATIONSHIP OF STEPS

Step 1 Determine symptoms	Step 2 Localize Trouble to functional unit	Step 3 Isolate Trouble to circuit	Step 4 Locate Specific trouble
Use senses make observations Knowledge of equipment Operate controls Record data	Step 1 plus use of: Diagrams Reasoning Bracketing Test equipment	Steps 1 and 2 plus use of: Schematic diagrams	Steps 1, 2, and 3 plus: Analysis Repair techniques Equipment checkout

THINGS TO DO DURING STEPS

FIGURE 1-4   Block diagram showing relationship between trouble-shooting steps

are made by noting the indications of the testing devices used to localize the trouble.

The third step, isolating the trouble to a circuit, uses all the information gathered up to this time. The main difference between this step and the second step is that it makes use of schematic diagrams instead of functional block diagrams. (It is also possible to use a detailed block diagram, known as a *servicing block diagram,* during this step.)

The fourth step, locating the trouble, involves review and verification of all the findings to make sure that the suspected part is the cause of the failure. The final step also includes the necessary repair (replacement of defective parts and so forth) and a final equipment checkout.

Example of relationship among troubleshooting steps. To make sure that you understand the relationship among troubleshooting steps, let us consider an example. Assume that you are troubleshooting a piece of equipment. You are well into the locate step of testing a suspected circuit, and you find *nothing* wrong with that circuit. That is, all wave-form, voltage, and resistance measurements within the circuit are normal. What is your next step?

You could assume that nothing is wrong, that the problem is *operator trouble.* This is absolutely *incorrect.* To begin with, there must be something wrong in the equipment because you recognized some abnormal symptoms before you got to the locate step. Never assume anything when troubleshooting electronic equipment. Either the equipment is working properly, or it is not working properly. Either observations and measurements are made, or they are not made. Either you draw valid conclusions from the observations, measurements, and other factual evidence, or you repeat the troubleshooting procedure.

Repeating the troubleshooting procedure. Some technicians new to service work assume that repeating the troubleshooting procedure means starting all over from the first step. In fact, some service literature recommends this action. The recommendation is based on the fact that it is possible for anyone, even an experienced technician, to make mistakes. And although the troubleshooting procedure will keep mistakes to a minimum if it is performed logically and systematically, voltage and resistance measurements can be erroneously interpreted, wave-form observations or bracketing can be incorrectly performed, and numerous other mistakes can be caused by simple oversight.

In spite of the recommendation by some writers of service literature, this author contends that repeating the troubleshooting procedure means *retracing your steps, one at a time, until you find the place where you went wrong.* Perhaps it was a previous voltage or resistance measurement erroneously interpreted in the locate step, or perhaps a wave-form observation or bracketing step was incorrectly performed in the isolate step. Whatever the cause, it must be found logically and systematically by taking a *return path* to determine *where* you went astray.

1-3. MODIFICATION OF THE TROUBLESHOOTING PROCEDURE

Thus far, we have established that a systematic, logical approach is absolutely essential for an efficient and successful troubleshooting procedure. Knowledge of the equipment, both theoretical and operational, is equally important. This knowledge will permit you to make decisions regarding possible modification of the troubleshooting procedure. Such modification may be necessary in order to apply the troubleshooting procedure intelligently to both simple and complex electronic equipment.

Obviously, it is impractical to perform the second step (localize the trouble to a functional unit) if there is only one unit (as is the case with a television set or a radio receiver). Similarly, it is not necessary to perform the third step (isolate the trouble to a circuit) if the equipment is a simple one-circuit device (such as a meter or a digital logic probe).

Figure 1-5 shows how the basic troubleshooting procedure can be modified or simplified, depending upon the design of the equipment. The following paragraphs summarize the modification procedure:

If the equipment contains more than one functional unit, you have no choice but to include all the steps of the troubleshooting procedure.

EQUIPMENT WITH MORE THAN ONE FUNCTIONAL UNIT

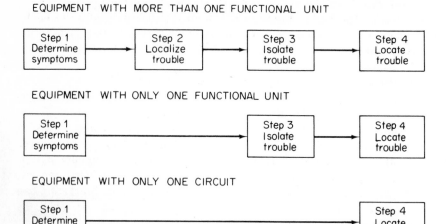

FIGURE 1-5 Block diagrams showing modification of trouble-shooting procedure

If there is only one functional unit and this unit contains more than one electronic circuit, you may omit the localize step and proceed directly from the symptoms step to the isolate step. Thus, you will now follow a *three-step* procedure.

If there is only one electronic circuit, you may omit both the localize and the isolate steps and proceed directly from the symptoms step to the locate step. Thus, you will now follow a *two-step* procedure.

1-4. DETERMINING TROUBLE SYMPTOMS

It is not difficult to realize that there is definitely trouble when a piece of electronic equipment will not operate. For example, there is obviously trouble when a television set is plugged in and turned on but there is no picture, no sound, and no pilot light. A different problem arises when the equipment is still operating but is not doing its job properly. Using the example of the television set, assume that the picture and sound are present but that the picture is weak and there is a buzz in the sound.

Another problem in determing trouble symptoms is improper use of the equipment by the operator. In the case of complex electronic equipment, such as a radar set, operators are usually trained and checked out on the equipment. The opposite is true of home entertainment equipment used by the general public. However, no matter what kind of equipment is involved, it is always possible for an operator to report a trouble that is actually the result of improper operation.

For these reasons, the you must *first* determine the signs of failure, *regardless* of the extent of malfunction, caused by either the equipment or the operator. This means that you must know how the equipment operates normally and how to operate the equipment controls.

1-4.1 Knowing the Equipment

All electronic equipment is designed to perform some function or group of functions, according to the requirements of the user and the equipment manufacturer. These requirements establish a certain type and level of performance that must be obtained at all times. Therefore, it would be impossible to maintain the equipment in top operating condition if you did not know when the equipment was not operating properly.

It is not necessary for you to know who is using the equipment, but it is important for you to know what the job of the equipment is and whether this job is being done. Similarly, it may be nice to know who manufactures the equipment, but this information is not of any great importance as far as troubleshooting is concerned.

A possible exception is troubleshooting specialized equipment such as television sets. For example, the adjustment controls of a Zenith color television set are different from those of an RCA set. Thus, the troubleshooting procedures for a Zenith-type set are different from those for an RCA-type set, even though the final results are the same. However, in general, you should be able to perform your troubleshooting task without knowing who the manufacturer is. The important questions to be considered at this time are: What is the specific job (such as reproducing sound, a picture, or a radar image) of the equipment, and is the equipment doing that job?

In order to achieve a fuller understanding of the equipment, it is important to remember that all electronic equipment, no matter how simple or how complex, is built by using various combinations and arrangements of electronic components and devices (resistors, capacitors, inductors, transformers, transistors, vacuum tubes, switches, and so forth). Basic electronic circuit theory describes the actions and effects of these components in their various arrangements. However, having a theoretical knowledge of the equipment is not enough; you must know how to apply this knowledge in a practical way. Service literature is written to aid you in applying your theoretical knowledge to the specific equipment in a practical manner.

To be a good troubleshooter, you must effectively combine theory and practical experience. Theoretical knowledge without practical experience (or vice versa) is almost useless. In troubleshooting, the theoretical and the practical are complementary. When you have only practical experience, you are limited because you may not understand what indications mean. For example, even if you monitored all wave forms in a particular circuit and compared the wave forms against those of the service literature, you still might not be able to understand why the actual wave form deviates from the desired wave form. Conversely, when you have only a thorough knowledge of electronic theory, even though you know the normal operation of the equipment, you are limited in troubleshooting because you cannot apply this knowledge in a practical manner. For example, if you cannot connect the test equipment and adjust the controls to make wave-form measurements, you will never be able to find a faulty wave form.

1-4.2 Use of Operating Controls during Troubleshooting

As a troubleshooter, you should not be required to operate equipment controls during normal performance conditions. However, you should be able to use the operating controls as well as, if not better than, the equipment operator. *Operating controls* are all *external* switches and controls connected through a shaft or other mechanical linkage to internal circuit components (variable resistors, potentiometers, variable capacitors, channel selectors, and so forth) that can be adjusted *without going inside* the equipment enclosure.

Operating controls are those controls that must be used in order to supply power to the equipment circuits (such as the ON-OFF switch), to tune or adjust the performance characteristics (such as the channel selector or volume control), or to select a particular type of performance (such as an AM-FM switch). By their nature, operating controls produce a change in the circuit conditions by the direct variation of resistance, capacitance, inductance, and so forth. This, in turn, causes an indirect change in the circuit current or voltage. The information displays of the equipment (front-panel meters, television picture tube, loudspeakers, and other audio or visual indicating devices) permit you to see or hear changes that take place when the operating controls are used.

Operating versus adjustment controls. Compare two controls such as the volume control of a radio receiver and the IF adjustment screw of the same receiver. Both controls tune or adjust the performance characteristics of the radio receiver. However, it is necessary to go inside the receiver to adjust the IF screw. Thus, the IF adjustment screw is definitely not an operating control. On the other hand, the volume control (probably a potentiometer) is a variable resistor having a shaft extending outside the equipment enclosure. Manipulation of this control causes a resistance change in a circuit in the audio section of the receiver, thereby varying the gain of the audio section. This, in turn, varies the sound or volume appearing at the loudspeaker. Thus, a volume control is definitely an operating control.

1-4.3 Improper Adjustment of Operating Controls

It is possible to damage some equipment by improper adjustment of operating controls. Every electronic circuit component has definite maximum current and voltage limits below which it must be operated in order to prevent burnout or insulation breakdown. The meters placed on the front panel of complex electronic equipment serve as aids in determining voltage and current values at critical areas in the equipment circuitry. Thus, it is important that the operating controls be adjusted so that the limits specified in the service literature for those areas are not exceeded.

Unless you observe the proper precautions while investigating the symptom, the improper use of operating controls can result in even more damage to the already defective equipment. Always refer to service literature when troubleshooting unfamiliar equipment, taking special note of any precautions. Observe the classic rule of electronics: When all else fails, follow instructions.

For example, the intensity control of an oscilloscope should never be adjusted to produce an excessively bright spot on the fluorescent screen because such a spot may burn the screen coating and decrease the life of the cathode-ray tube. A knowledge of the circuit changes that take place when you adjust a control will enable you to think ahead of each step and thus anticipate any damage that the adjustment might produce. Never make an adjustment in haste or in panic.

1-4.4 Initial Settings of Operating Controls

With all this discussion concerning proper use of operating controls during troubleshooting, you may wonder what the best settings for controls are *before power is applied.* It is obvious that the controls should not be set to the maximum positions. This will most certainly result in improper operation and will probably cause damage. On the other hand, setting all controls to minimum positions is not the best approach (unless you are specifically directed to do so by the service literature). Minimum settings could cause more trouble than might be imagined, especially if the controls are used *in the wrong order.*

For example, to prevent damage to the output stages of high-powered vacuum-tube radio transmitters, it is essential that the filament voltage be applied first for filament warm-up *before* the high plate voltage is applied. When the transmitter is turned off, the plate voltage is removed first; then, the filament voltage is removed. If (by accident or through ignorance of the equipment) the filaments are turned off before the plate voltage is removed, damage to the tubes may result. A similar situation can occur in some equipment if the bias is removed or decreased to a low value while the plate voltage is at a high value. Clearly, setting all controls to the OFF or minimum position is not the answer.

Generally, service literature recommends that all operating controls be set to a *safe* or *normal* position. Of course, these positions are not always the same for all types of electronic equipment or even for two different pieces of the same type of equipment. The service literature must be consulted in all cases.

By setting all controls to their safe or normal positions as an initial step in troubleshooting, you will gain information that can define the trouble symptoms still further and aid you in proceeding with your trouble analysis. For example, if all the controls are set at their correct positions but the symptom persists, it is possible that an operating control is responsible for the trouble symptom. In this case, the trouble would have to fall within the area of component failure. If a control is faulty, this may be immediately apparent, especially when it is a mechanical failure. For example, if a volume-control potentiometer has a bad spot on the resistance element, moving the control to the correct operating position will result in indications (scratching sounds) on the loudspeaker. Of course, additional information may be required to determine when a control has failed electronically because the trouble symptom produced may also point to other electronic failures.

To sum up, if the trouble symptom is the result of a control-setting error, it is possible to remedy the trouble immediately by setting the controls to their proper position (as an initial step in troubleshooting). If the trouble is the result of control failure, proper adjustment of the controls will aid in locating the defective control. Even if the trouble is not directly related to the controls, the resultant effect of controls on the symptom may aid in locating the trouble.

1-4.5 Recognizing Trouble Symptoms

Symptom recognition is the art of identifying the normal and abnormal signs of operation in electronic equipment. A trouble symptom is an *undesirable change* in equipment performance or a deviation from the standard. For example, the normal television picture is a clear, properly contrasted representation of an actual scene. The picture should be centered within the vertical and horizontal boundaries of the screen. If the picture should suddenly begin to roll vertically, you should recognize this as a trouble symptom because it does not correspond to the normal performance that is expected. If you cannot recognize this as a trouble symptom, your technical ability to align a television receiver or to perform any repair procedure on the set will be useless.

1-4.6 Equipment Failure versus Degraded Performance

There are two broad categories into which trouble symptoms of abnormal equipment performance can be divided: *equipment failure* and *degraded performance*.

Equipment failure means that either the entire piece of equipment or some functional part of it is not operating properly. For example, the total absence of sound from a radio receiver when all controls are in their proper positions indicates a complete or partial equipment failure. Similarly, the absence of a picture on the screen of a television set when all controls are properly set is a form of equipment failure, even though there may be sound from the television loudspeaker.

Degraded performance occurs whenever the equipment is working but is presenting the operator with information that does not correspond to the design specifications or to normal performance. This degraded performance may range from nearly perfect to barely operating. For example, the presence of hum in the output of a radio receiver is degraded performance because the equipment has not yet failed but the performance is abnormal.

Example of equipment failure and degraded performance. Generally, the terms *equipment failure* and *degraded performance* apply to the equipment's built-in indicators. However, with some types of equipment, it may be necessary to use test equipment to distinguish between the two conditions. For example, assume that the equipment is a simple, one-stage amplifier like the one shown in Fig. 1-6. Using the service literature or your knowledge of similar circuits, you know that if a sine wave is applied to the transistor base (input), an amplified sine wave should appear at the collector (output). The amount of amplification will depend upon the gain of the circuit. Assume that the input sine wave is 1 volt (V) and that the supposed gain is 10. Now, let us analyze each of the four possible wave forms shown in Fig. 1-6.

Wave form *A* is an example of equipment failure. With a normal wave form at the base (input), there is no wave form at the collector (output).

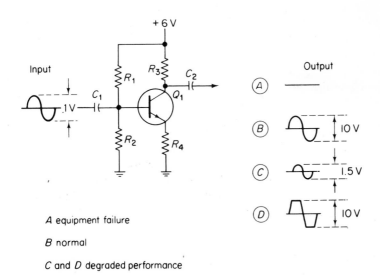

A equipment failure

B normal

C and *D* degraded performance

FIGURE 1-6   Examples of equipment failure and degraded performance

Assuming that there are no operating controls and that power is applied, wave form *A* indicates that the amplifier has failed to perform its function completely. (The next troubleshooting step would be substitution of parts and/or voltage and resistance checks, as discussed in later sections of this book.)

Wave form *B* indicates neither equipment failure nor degraded performance. That is, the collector wave form (output) is a sine wave approximately 10 times greater than the base wave form (input). This is the normal indication, and no trouble exists.

Wave form *C* is an example of degraded performance. There is a wave form at the collector (output), so the amplifier circuit is performing part of its function. However, the output wave form is not much greater in amplitude than the input wave form. Thus, the amplifier is producing little or no gain (in any event, the gain is well below the desired ten).

Wave form *D* is also an example of degraded performance. There is a wave form at the output, and it shows the desired gain of ten. However, the output wave form is distorted. That is, the output wave form is not an amplified version of the input wave form, which it should be in a properly functioning circuit.

Note that the processes used here to obtain the wave forms are examples of signal substitution and signal tracing. Both techniques are discussed in great detail throughout this book.

1-4.7 Evaluation of Symptoms

Except for the very simple type of equipment used in the previous example, it is generally not realistic to use test equipment immediately for recognition of a trouble symptom. Instead (in most cases), an evaluation of the symptoms is made long before test equipment is used.

Symptom evaluation is the process of obtaining more detailed descriptions of the trouble symptoms. The purpose of symptom evaluation is to enable you to understand fully what the symptoms are and what they *truly* indicate so that you will be able to gain further insight into the problem. Unless you first define a trouble symptom completely, you can quickly and easily be led astray. The result could be a loss of time, unnecessary expenditure of energy, or perhaps even a total dead-end approach.

The recognition of the original trouble symptom does not in itself provide enough information for you to decide on the probable cause or causes of the trouble. The reason for this is that many faults produce *similar trouble symptoms*. In order to evaluate a trouble symptom, you generally have to use the operating controls associated with the symptom and apply your knowledge of basic electronics, supplemented with information gained from the service literature.

Of course, the mere adjustment of the operating controls to their normal positions is not the complete story of symptom evaluation. However, the discovery of an incorrect control setting can be considered a part of the overall symptom-evaluation process.

Example of evaluating symptoms. When the screen of a television set is not on (no raster), there is obviously trouble. The trouble could be caused by the brightness control being turned down (assuming that the power cord is plugged in and the ON-OFF switch is set to ON). However, the same symptom can be produced by a burned-out picture tube or a failure of the high-voltage power supply, among many other possible causes. Think of all the time you may save if you check the operating controls first, before you charge into the set with tools and test equipment.

It is not necessary, and usually not practical, to list all the possible causes of the original trouble symptom in order to evaluate the symptom properly. For example, recognizing an undesirable hum in a radio receiver as a trouble symptom could lead you in several directions if you do not obtain a more detailed description of the symptom. The hum may be caused by poor filtering in the power supply circuits, by ac line-voltage interference, or by heater-cathode leakage (in vacuum-tube sets), to name just a few causes.

To be effective, you must make a valid decision concerning the *most probable* cause of the symptom. Recognizing the original symptom does not usually provide sufficient information for making this decision; evaluating and elaborating on the symptom will provide that information.

1-4.8 Recording Troubleshooting Information

Trouble symptoms must be evaluated in relation to *one another*, as well as in relation to the overall operation of the equipment. One of the easiest ways of making this evaluation is to have *all* data handy for reference by recording the information as it is obtained.

Recorded data will enable you to sit back and think the information over before coming (rather than jumping) to a conclusion about the source of the trouble. Properly recorded data will also enable you to check the service literature and compare your information with detailed descriptions if this is necessary.

By recording all control positions and the associated indicator information, you can quickly evaluate the information and check to see that it is correct. You can also use these data to put the equipment in exactly the operating condition that you want to test.

Keep in mind that it is possible to set controls incorrectly, to operate controls in an incorrect order, and to misread meters and other indicators. Thus, it is sometimes necessary to reproduce certain operating conditions or to perform additional tests in order to check your recorded information. In addition, whenever the adjustment of a control has no effect upon the symptoms, that fact should be recorded. Such information may later prove to be just as important as any changes a control may produce in the trouble symptom. For example, the fact that a volume control has no effect on volume is just as important as the fact that the control must be set to full ON to get any volume.

1-4.9 Evaluating Equipment Performance

It is obvious that a knowledge of the normal equipment displays will enable you to recognize abnormal displays. It should also be obvious by now that during the process of doing its assigned job, most electronic equipment yields information that an operator or a technician can either see or hear.

Electrical information to be presented as sound can or should be applied to a loudspeaker or a headset. A visual display results when the electrical information is applied to a cathode-ray tube or to an indicating meter that is built into the equipment's control panel and can be viewed by the operator. In the case of digital equipment, the register lights indicate functions being performed by internal circuits. In some equipment, individual circuits or plug-in modules have indicators that show whether the circuit or module is working properly. Even pilot lights can provide a visual indication of equipment operation.

The senses of hearing and sight allow you to recognize the symptoms of normal and abnormal equipment operation and thus help you to evaluate the performance of the equipment. Keep in mind that the display of information is generally the sole job of the equipment but that this display may also perform the secondary job of providing information to permit performance evaluation.

During performance evaluation, the operating controls must be manipulated in order to evaluate the equipment performance. However, this is done only in a predetermined and knowledgeable manner, with due regard for proper operation of the equipment. For example, you do not operate the equipment controls until the displays correspond to what you think they should be for a particular mode. At *no* time during the troubleshooting procedure should controls be operated arbitrarily or indiscriminately.

When evaluating equipment performance, pay close attention to the displays that the *equipment itself* produces, and compare these displays with your knowledge of how the equipment should normally perform. Generally, performance evaluation is accomplished without the use of any test equipment whatsoever. (The exception to this rule is simple equipment that has no displays, such as the amplifier described in Sec. 1-4.6). A careful study of the built-in indicators will usually provide sufficient evaluation information to enable you to recognize a trouble symptom.

1-4.10 Summary of Determining Trouble Symptoms

To do a truly first-rate job of determining trouble symptoms, you must have a complete and thorough knowledge of the normal operating characteristics of the equipment. Your knowledge helps you to decide whether the equipment is doing the job for which it was designed. In most service literature, this is more properly classified as "knowing your equipment."

In addition to knowing the equipment, you must be able to operate properly all the controls associated with the equipment in order to determine the symptoms, that is, in order to decide whether the equipment is performing normally or abnormally. If the trouble is cleared up by manipulating the controls, your trouble analysis may or may not stop at this point. By using your knowledge of the equipment, you should be able to find the reason *why* the specific control adjustment removed the apparent trouble.

When adjusting controls, use extreme caution. A misadjustment may cause additional circuit damage. Also, you must be aware of the *circuit area* in which the control is located. Only those controls that will logically affect the indicated system should be adjusted. Gaining further information about a trouble symptom by properly setting the operating controls will help you localize the trouble to a functional unit, which is the next step in the troubleshooting procedure.

1-5. LOCALIZING TROUBLE TO A FUNCTIONAL UNIT

To localize the trouble to a functional unit, you make use of symptom recognition, elaboration, and evaluation to determine which functional units are the *probable* source of the trouble. You then determine which of the selected functional units is the *actual* source of the trouble by using test equipment to check their *inputs* and *outputs* in a logical, systematic manner.

After completing this step, you verify that you have found the functional unit containing the trouble and make sure that the unit is malfunctioning and that it is the *only* one that is not operating properly. However, the completion of this step will *not* isolate the trouble to a defective component; that comes later in the troubleshooting process.

1-5.1 The Meaning of "Localize" in Troubleshooting

Localizing the trouble to a functional unit means that you must determine which of the major functional areas (or units) of a multifunctioning piece of equipment is actually at fault. This is accomplished by systematically checking each unit selected until the actual faulty one is found. If none of the functional units on your list show improper performance, you must take the return path and recheck the symptom information (and gather more information if possible). There may be several circuits and/or components that could be causing the trouble. The localize step will narrow the list to those in one functional area, as indicated by a particular block in the equipment's functional block diagram.

1-5.2 Functional Divisions of Equipment

We have defined a function as an electronic operation performed in a specific area of a piece of equipment. The combined functions cause the equipment to perform the overall electronic purpose for which it was designed. Frequently, the terms *function* (an operational subdivision) and *unit* (a physical subdivision) are used interchangeably.

Figure 1-7 shows the relationships of the functional divisions of a piece of equipment and the four basic troubleshooting steps.

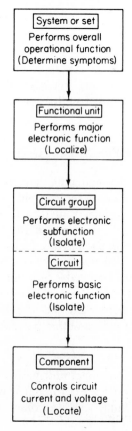

FIGURE 1-7 Relationship of divisions of system or set and basic troubleshooting steps

A *system* or *set* (radar system, television set, communications set, and so forth) is designed to perform an overall operational function. The *determine-symptoms* step of the basic troubleshooting procedures is associated with the system or set classification.

The system or set is divided into *functional units* (transmitter, receiver, and so forth), each designed to perform a major electronic function vital to the overall operational function. The *localize* step is associated with the functional-unit division. Note that when there is only one functional unit, the localize step is omitted, as shown in Fig. 1-8.

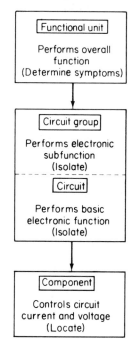

FIGURE 1-8 Relationship of divisions of functional unit and basic troubleshooting steps

The *circuit groups* (RF section, IF section, and so forth) or the individual *circuits* (amplifier, oscillator, and so forth) are subdivisions of the functional unit and perform electronic subfunctions. In some equipment, there is only one circuit in a circuit group (e.g., only one IF amplifier). In other equipment, there is more than one circuit (often many) in a circuit group (e.g., an IF section that has three amplifier circuits). The *isolate* step of the troubleshooting procedure is concerned with finding which circuit is at fault.

The final subdivision of electronic equipment is the *component* or *part*. These components (transistors, resistors, capacitors, and so forth) control the circuit current and voltage. The *locate* step of the troubleshooting procedure is associated with the component or part subdivision except in the case of equipment that is composed primarily of integrated circuits (ICs).

The troubleshooting sequence must be modified when ICs are involved, as shown in Fig. 1-9. The last step (locate defect to a part) is unnecessary (if not impossible to accomplish). ICs are made up of many parts to form one circuit or several circuits. They are sealed and are replaced as a package. Thus, once trouble is isolated to the IC, the troubleshooting sequence is complete except for repair and checkout. For example, if an IC is used to form the complete IF section of a television set and trouble has definitely been isolated to the IF section, the next logical step is to replace the IC and check the set.

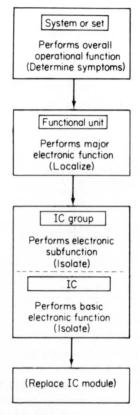

FIGURE 1-9 Relationship of divisions of system or set (with ICs) and basic troubleshooting steps

1-5.3 Functional Block Diagrams

A functional block diagram is an overall representation of the functional units within the equipment, as well as of the *signal-flow paths* between units. Figure 1-10 shows a typical functional block diagram for an amplitude-modulated transceiver set (transmitter and receiver) composed of six func-

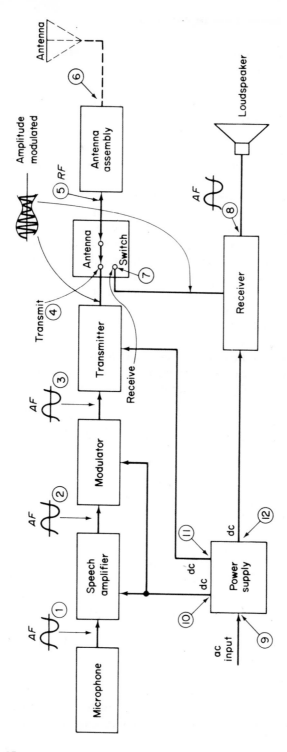

FIGURE 1-10 Functional (overall) block diagram of transceiver

25

tional units. In some service literature, such an illustration is referred to as an *overall block diagram*. Remember that the functional units may be physically separate, may be located in different sections of a single chassis, or may be separate plug-in modules mounted on a common chassis.

In Fig. 1-10, there is no indication of how each function is accomplished. Thus, each unit may consist of a variety of circuits or stages, each performing its own major electronic function. For example, the receiver unit may contain RF, IF, and audio stages, as well as a local oscillator. However, the major electronic function of the receiver is to convert the RF carrier wave into audio signals that are reproduced on the loudspeaker.

Notice that the connecting lines between the various functional blocks (or units) represent *important signal-flow connections* but that the diagram does not necessarily show where these connections can be found in the actual equipment circuitry. In some cases, additional information such as frequencies and input-output wave forms may be included to show how far each type of signal progresses through the equipment. The layout interpretation of the connecting lines and the representation of the blocks will depend upon the complexity of the equipment and the care with which the service literature is written.

Interpreting the functional block diagram. A functional block diagram does not show physical relationships because there is no relationship whatsoever between the physical location of the units within the equipment and the block arrangement in the functional block diagram. However, the diagram does provide a general representation of the major functional units and the *important signal relationships among them*. Thus, the block diagram can help you to understand the equipment's overall purpose. That is, it can help you to answer the question: What major functions must be accomplished to perform the overall task for which the equipment is designed?

In most service literature, the overall block diagram is supplemented with written descriptions of the *overall theory of operation*. For example, if the diagram of Fig. 1-10 were used in a typical technical manual, the supporting theory of operation would be something like this: "During transmission, the *microphone* converts the sound information to be transmitted into an electrical signal of audio frequency (AF). The *speech amplifier* amplifies the AF signal and applies it to the *modulator*, which, in turn, applies the same signal to the *transmitter* in a manner that causes the amplitude of the RF carrier signal to vary at an audio rate.

"In addition to providing the RF carrier signal, the *transmitter* amplifies the signal to a suitable level. The *antenna assembly* converts the RF signal into electromagnetic energy. The antenna assembly consists of an antenna and the lead-in cable to the transceiver.

"When the transceiver is serving as a receiver, the antenna assembly picks up the incoming RF signal and applies it to the *receiver* through the *antenna switch.* The receiver removes any AF information from the signal and applies the AF portion of the signal to the loudspeaker.

"The *power supply* converts the ac line voltage into dc voltages suitable for operation of the various units and circuits. The *antenna switch* connects the antenna to the transmitter output or receiver input, depending on the mode of operation (transmit or receive)."

1-5.4 Bracketing Technique

The basic bracketing technique makes use of the functional block diagram to localize the trouble to a functional unit. Bracketing (sometimes known as the *good-input–bad-output technique*) provides a means of narrowing the trouble area down to a single faulty functioning unit, then to a circuit group, and, finally, to a faulty circuit. Symptom analysis and/or signal-tracing tests are part of bracketing or are used in conjunction with it.

Bracketing begins with the placement of brackets on the block diagram at the good input and the bad output. Bracketing can also be used on the schematic diagram in the case of simple equipment. The brackets can be in your mind or physically marked on the diagram with a pencil, whichever is most effective for you. No matter what system you use, if the brackets are properly positioned, you will know that the trouble exists somewhere *between* the two brackets.

The technique is to move the brackets one at a time (either the good input or the bad output) and then make tests to determine whether the trouble is within the new bracketed area. The process continues until the brackets localize a single defective unit.

The most important factor in bracketing is to decide *where* the brackets should be moved in the elimination process. This is determined from your deductions based on your *knowledge of the equipment* and on the *symptoms.* All moves of the brackets should be aimed at localizing the trouble with a *minimum* of tests.

1-5.5 Examples of Bracketing

Depending upon the type of equipment involved, bracketing can be used *with* or *without* actual measurements of signals or wave forms at test points. That is, localization can sometimes be accomplished on the basis of symptom evaluation alone. In practical troubleshooting, both symptom evaluation and tests must be made, often simultaneously. The following examples show how the technique is used in both cases. These examples assume that a citizens band (CB) transceiver is being serviced and that a functional block diagram (with corresponding test points) similar to Fig. 1-10 is provided in the service literature.

Note that some test points are both input and output for different units. For example, test point 3 is the input to the transmitter and the output of the modulator. Test point 3 is also the final test point in the AF portion of the transmission system. The signal at test point 3 is audio frequency and corresponds to whatever is applied at the microphone (such as a voice). In the bracketing system, test point 3 could be a *good input* to the transmitter or a *bad output* from the modulator (or a bad output from the entire AF portion of the transmission system).

Test point 1 is both an output from the microphone and an input to the speech amplifier. Test point 2 is an output from the speech amplifier and an input to the modulator. Test points 1, 2, 3, and 8 monitor AF signals or voltages and must be tested with an oscilloscope or ac voltmeter. In an emergency, these test points could be monitored with a headset or loudspeaker.

Test points 10, 11, and 12 could be considered inputs to the various functional units. However, these test points are dc voltages, rather than signal voltages, and as a practical matter should be considered outputs from the power supply. Test point 9 monitors the ac input to the power supply. Test points 9 to 12 must be monitored with a voltmeter.

Test points 4, 5, 6, and 7 monitor RF voltages, possibly modulated with AF signals. Thus, an oscilloscope or voltmeter with an *RF probe* is required. If the AF voltage present on the RF carrier is to be monitored during troubleshooting, a *demodulator probe* must be used. Probes, oscilloscopes, meters, and all other test equipment associated with troubleshooting are discussed in Chapter 2. For the following discussion, it is sufficient for you to understand that different types of test equipment are required at different test points.

Example of no-reception trouble symptom. Assume that the operator reports proper transmission but no reception. You confirm the symptom by operating the equipment through its normal sequence. The receiver's channel selector and volume control have no effect on the symptom; with the controls set to RECEIVE and the volume control to full ON, there is no sound from the loudspeaker. A known-good transmitter is operated nearby (on the appropriate channel), but there is no voice or other signal on the loudspeaker.

On the basis of the trouble symptom and using bracketing, it is possible to eliminate several functioning units without using any test equipment. For example, because the problem is in reception, the microphone, speech amplifier, modulator, and transmitter can be eliminated. However, the receiver, antenna switch, antenna assembly, and power supply could be at fault. Thus, each of these units should have brackets placed around it, as shown in Fig. 1-11.

The next step is to eliminate as many units as possible (ideally, three out of the four) by test or substitution. The trouble is then localized to the one remaining unit. The localization procedure can take one of two courses, depending upon the design of the equipment.

Plug-in modules. If the functional units are plug-in modules, which is the case in much solid-state equipment, each of the four modules can be

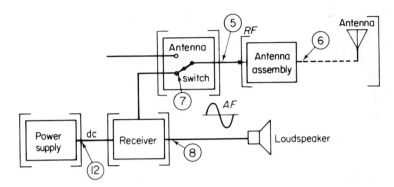

FIGURE 1-11   Placement of brackets for no-reception trouble symptom

replaced in turn until the trouble is cleared. For example, if replacement of the receiver module restores normal operation, the defect is in the receiver. This can be confirmed by plugging the suspected defective module back into the equipment. (Although this confirmation process is not a necessary part of *theoretical* troubleshooting, it is wise to make the check from a *practical* standpoint. Often, a trouble symptom of this sort can be caused by the plug-in module making poor contact with the chassis's connector or receptacle.)

Equipment without plug-in modules. If the functional units are not the plug-in type, or if they are not readily replaceable, localization *must* be accomplished by means of tests. This is also true if replacement plug-in modules are not readily available in the field.

In the case of the power supply, the dc voltage (at test point 12) to the receiver should be measured at both the power supply and the receiver ends of the wiring. If the dc voltage is correct at the receiver, the power supply can be eliminated. If the dc voltage is correct at the power supply but not at the receiver, the wiring between is the most logical suspect. (The wiring could be broken; terminals could be making poor contact; and so forth.)

With the trouble narrowed down to the receiver, antenna switch, or antenna assembly, the next step is to decide whether signal injection or signal tracing will permit the fastest localization. (Both signal injection and signal tracing are discussed in greater detail in Sec. 1-6 and throughout this book.)

Signal injection. If signal injection is used, a signal generator is tuned to the operating frequency of the receiver radio frequency and modulated by an AF tone, such as 1 kilohertz (kHz). The signal-generator output is injected at test points 7, 5, and 6 (in that order), with the transceiver controls in the receive position of operation, as shown in Fig. 1-12. If a tone is heard in the loudspeaker with a signal at test point 7, the receiver can be eliminated, and the trouble is most likely in the antenna switch or antenna

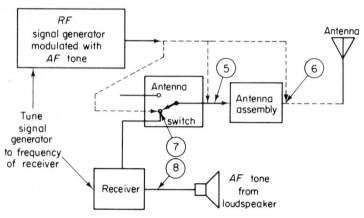

TRANSCEIVER CONTROLS IN RECEIVE POSITION

FIGURE 1-12   Troubleshooting no-reception trouble symptom with signal injection

assembly. If a tone is heard with a signal at 5 but not at 6, the antenna assembly is the most likely cause of trouble (perhaps there is a broken antenna connector). Keep in mind that if transmission is normal, the antenna assembly is probably good, and the problem is in the antenna switch.

Signal tracing. If signal tracing is used, a signal generator is tuned to the operating frequency of the receiver and is modulated by an AF tone. The signal-generator output is connected to test point 6, and the signal is traced (by means of oscilloscope or meter with RF and demodulator probes) at points 6, 5, and 7 (in that order), as shown in Fig. 1-13. If the signal is present at 5 but not at 7, the antenna switch is defective. If the signal is present at 7 but there is no tone from the loudspeaker, the trouble is localized to the receiver.

Another example of no-reception trouble symptom. Assume that the operator again reports no reception, which you confirm. However, you now find that there is a scratching sound from the loudspeaker when you adjust the volume control. You also check the transmitter and find that all front-panel meters provide normal indications. In the absence of panel meters for the transmitter, you confirm normal transmission from the fact you can be heard on a nearby CB station.

In this case, the trouble-localization process is similar to that described for the previous no-reception example except that *in theory*, the power supply can be eliminated immediately without further tests. (From a practical troubleshooting viewpoint, the more units that can be eliminated from suspicion without testing, the better.)

The reasoning for elimination of the power supply as the trouble source is as follows: If all circuits of the power supply are defective, the transmitter

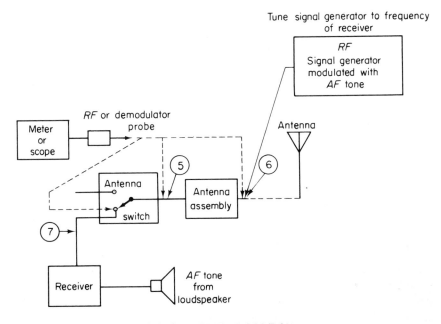

TRANSCEIVER CONTROLS IN RECEIVE POSITION

FIGURE 1-13 Troubleshooting no-reception trouble symptom with signal tracing

function will not be present, nor will the panel meters provide normal indications. If the power supply is functioning partially (voltages to all units except the receiver), the receiver will be dead, and there will be no sound from the loudspeaker when the volume control is operated.

On the other hand, it is not absolutely certain that the power supply has been cleared of possible fault. Assume, for example, that the power supply is delivering correct voltages to all units except the receiver because of a defect in the wiring between the power supply and the receiver. Or assume that the receiver voltage comes from a separate circuit within the power supply and that only that particular circuit is defective and is delivering a low voltage.

These points are made to demonstrate the difference between theoretical and practical troubleshooting. In theory, with the symptoms described, the power supply is eliminated from any possibility of being a defective unit. In practical troubleshooting, the same symptoms localize trouble to the receiver, antenna assembly, and antenna switch as the *most likely* areas of trouble and as the *starting points* for further trouble localization. Thus, each of these units should have brackets placed around them, as shown in Fig. 1-14.

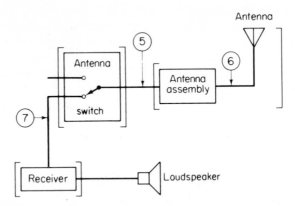

FIGURE 1-14 Placement of brackets for no-reception trouble symptom (but with some sound on loudspeaker)

The next step is to eliminate as many units as possible (ideally two out of the three) by test or substitution. The trouble is then localized to the one remaining unit. The localization procedure is essentially the same as the one used for the previous no-reception trouble: substitution of plug-in modules, signal injection, and/or signal tracing at test points 5, 6, and 7.

Example of no-transmission trouble symptom. Assume that reception is normal on all channels of the CB transceiver but that there is no transmission on any channel. You confirm that it is impossible to contact anyone on any channel, and you note that the transmitter's tuning-meter indication is low but that the modulation-meter indication is correct.

The *most likely* defective functional units are the transmitter, antenna switch, and antenna assembly. Brackets should be placed around these units, as shown in Fig. 1-15. The remaining units can be eliminated from the localization process by means of the following logic: Reception is normal; therefore, the receiver is operating properly. There is a normal indication on the modulation meter; therefore, the microphone, speech amplifier, and modulator are in good condition. If the power supply is defective, none of

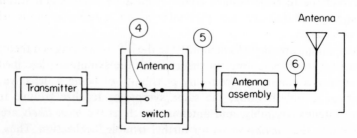

FIGURE 1-15 Placement of brackets for no-transmission trouble symptom

the functions will be correct. Of course, if the power supply is delivering a low voltage to the transmitter but correct voltages to all other units, the same symptoms will occur. However, this is *not* a most likely condition.

If the transmitter, antenna switch, and antenna assembly are the plug-in type, they can be replaced one at a time until the trouble is cleared. If the units are not readily replaceable, tests must be made to localize the trouble. Note that the antenna assembly is the least likely cause of trouble because the antenna is used for both transmission and reception (and reception is normal in this example). However, it is possible that the antenna could have some defect that will show up only during transmission (e.g., poor insulation in the antenna cable that arcs when the transmitter is used).

Because a transmitter provides or generates its own signal, signal tracing is the most logical test method (in preference to signal injection). The transceiver is operated in the TRANSMIT mode, and the signal is traced (by means of an oscilloscope or meter with RF and demodulator probes) at points 4, 5, and 6 (in that order), as shown in Fig. 1-16.

If the signal is absent at test point 4, the transmitter unit is at fault. If the signal is present at 4 but not at 5, the antenna switch is defective. A good signal at 5 but no signal at 6 indicates a defective antenna assembly.

1-5.6 Which Unit to Test First

When functional units of a piece of equipment are not readily replaceable and you have localized trouble to a group of units, you must decide which unit to test first. Several factors should be considered in making this decision.

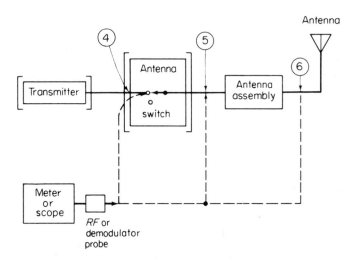

TRANSCEIVER CONTROLS IN TRANSMIT POSITION

FIGURE 1-16 Troubleshooting no-transmission trouble symptom with signal tracing

As a general rule, if you can make a test that eliminates *several* units, you should choose that test in preference to a test that eliminates only one unit. This decision requires an examination of the block diagram and a knowledge of how the equipment operates. The decision also requires that you apply logic in making your selection.

Test-point accessibility is the next factor to consider. A test point can be a special jack located at an accessible spot on the equipment, such as the front panel or the chassis. The jack (or possibly a terminal) is electrically connected (directly or by a switch) to some important operating voltage or signal path. At the other extreme, a test point can be any point where wires join or where parts are connected.

Other factors (although definitely not the most important) are past experience and a history of *repeated equipment failures*. Past experience with identical or similar equipment and related trouble symptoms, as well as the probability of failure based upon records of repeated failures, should have some bearing on the choice of a first test point. However, you should test all units related to the trouble symptom, no matter how much experience you may have with the equipment. Of course, experience can help you decide which unit to test first.

Example of choosing the first test point. Assume that the transceiver shown in Fig. 1-10 is being serviced. There is no audible signal from the receiver, and no effect is produced by manipulating the receiver's volume control. These are the same symptoms described in Sec. 1-5.5 for a no-reception trouble.

Assume that all test points on the transceiver are equally accessible. Further assume that the receiver has a previous history of failure in the local-oscillator section and that you decide to test by signal injection and voltage measurement.

Under these conditions, trouble is localized to the power supply, receiver, antenna switch, or antenna assembly. These units must be bracketed, as shown in Fig. 1-17, which also shows the test points involved (5, 6, 7, 8, 12) and the most logical order in which the points should be used. The following discussion illustrates the reasoning used to select this test-point order.

Either test point 7 or test point 12 is a good choice for the first check. If there is a good response to a signal injected at 7, both the power supply and the receiver are cleared of suspicion and the trouble is localized to the antenna or antenna switch. If response is bad with a signal at 7, the trouble is localized to the receiver or power supply. A check of the voltage at point 12 will then isolate the trouble to one unit; if the voltage at 12 is correct, the trouble is most likely in the receiver.

If the receiver shows some sign of operation (background noise from the loudspeaker that is adjustable by means of the volume control, but no signal) and all other conditions are identical, the most logical first test point is 7 rather than 12. The fact that the volume control will adjust the noise level

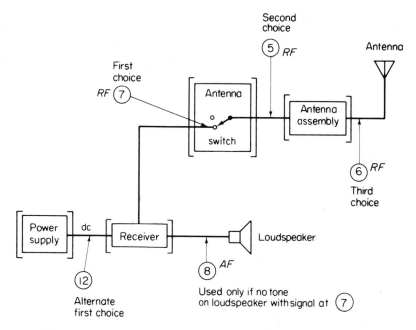

FIGURE 1-17 Example of test-point sequence using signal injection to troubleshoot a no-reception symptom

indicates that at least the audio stages are operating properly. Because these circuits require voltages from the power supply, it is safe to assume that the power supply output is satisfactory. This fact, plus the history of previous local-oscillator failure, leads you to conclude (tentatively) that the receiver is at fault.

Test point 8 can be eliminated as a first test choice because a signal injected at 8 will prove only whether the loudspeaker is good or bad. Test point 6 can also be eliminated as a first choice. If there is no response to a signal injected at 6, this proves nothing and only confirms the basic no-reception trouble symptom. Test point 5 is also a poor first choice. If there is no response to a signal injected at 5, this proves only that the antenna assembly is *probably good* but does not prove that the antenna switch is either good or bad.

Keep in mind that although test points 5, 6, and 8 are poor first test choices compared with test points 7 and 12, they should not be eliminated from the trouble-localization process. For example, if a signal injected at 7 shows that the receiver and loudspeaker are operating normally, the *next* logical points for signal injection would be 5 and 6.

Another example of choosing the first test point. Assume, once again, that the transceiver shown in Fig. 1-10 is being serviced. In this case, reception is good on all channels, but transmission is erratic. The transmitter tuning-meter indication is low, and the modulation-meter reading is erratic. Assume that all test points are equally accessible and that there is no previous history of similar troubles. You decide to test by signal tracing and voltage measurement (signal tracing is logical for a transmitter, as previously discussed).

Under these conditions, trouble is localized to the power supply, speech amplifier, modulator, and transmitter. These units must be bracketed, as shown in Fig. 1-18, which also shows the test points involved (1, 2, 3, 4, 10, 11) and the most logical order in which the points should be used. The following discussion illustrates the reasoning used to select this test-point order.

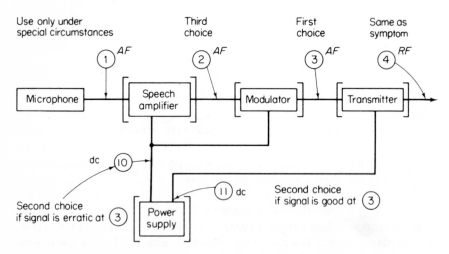

FIGURE 1-18 Example of test-point sequence using signal tracing to troubleshoot an erratic-transmission symptom

Test point 3 is the best choice for the first test. If the signal is erratic, trouble is localized to the speech amplifier, modulator, or one section of the power supply. A good signal at 3 will localize the trouble to the other power supply section or the transmitter. *Always aim for simultaneous elimination of several functional units* whenever accessibility of test points permits.

Now, assume that an erratic signal is found at 3. The next most logical step is to measure the voltage at 10. If the voltage is abnormal, the problem is in the power supply. If the voltage is normal, the trouble is in the speech amplifier or modulator (or possibly the microphone). A further test of the signals at 1 and 2 will confirm which unit is at fault.

Test points 2, 10, and 11 are better choices for the first test than 1 and 4, but not so good as 3. An erratic signal at 2 will prove a defect only in the speech amplifier (or possibly the microphone) or in one section of the power supply. A bad voltage reading at 10 or 11 localizes trouble to one section of the power supply.

From a theoretical standpoint, point 1 can be eliminated as a first test choice because it will only prove whether the microphone is good or bad. However, from a practical standpoint, test point 1 could be used as the first test choice if the microphone is the plug-in type and is readily accessible and a known-good microphone is available. Under these circumstances, substitute the good microphone, and check the equipment's performance.

Test point 4 can be eliminated as a first test choice under all circumstances. An erratic signal at 4 will prove nothing. This is the same as the basic symptom of erratic transmission.

1-5.7 Modifying the Trouble-Localization Sequence

Anyone who has had any practical troubleshooting experience knows that all the steps in a localization sequence can rarely proceed in textbook fashion. Just as true is the fact that troubles listed in equipment service literature very often *never* occur in actual use. These troubles are described in the literature to serve as a guide, not as an all inclusive list. In some cases, it may be necessary to modify your troubleshooting procedure when localizing the trouble. The physical arrangement of a system may pose special troubleshooting problems. On the other hand, experience with similar equipment may provide you with special knowledge that can simplify the localizing procedure.

For example, assume that the transceiver shown in Fig. 1-10 is being serviced. Transmission and reception are weak on all channels. The modulation-meter indication is normal, but the transmitter-meter indication is low. All stations can be received, and the volume control works, but all signals are weak.

With these symptoms, the power supply and the antenna assembly are logical choices. Assume that the transceiver's service literature lists both units as "probable cause" for failure (with these symptoms) because both units are common to transmission and reception. The antenna assembly is a more logical selection than the power supply because the modulator meter shows a normal indication. However, there may be a practical problem to consider.

Although the antenna is a logical first choice, it may be advisable to test the power supply unit first if the antenna assembly is not readily accessible or if the antenna-assembly tests are more difficult to make. For example, assume that the antenna is mounted at some remote location, that the antenna lead-in is long, or that there is an antenna-loading network located inside the equipment but that the power supply voltages can be measured at the front panel. Obviously, the power supply should be checked first.

Now, suppose that both the antenna assembly and the power supply check out satisfactorily (contrary to what you expected or were led to believe by the information in the service literature). At some point in making your selection, you may have overlooked some data in the symptom evaluation or missed a possible faulty unit. At this stage, panic may be setting in!

Under these circumstances, your first thought might be to reconsider the original symptoms or possibly to return immediately to the symptom-evaluation phase of localization. But (as previously discussed), you will do better to *retrace* your steps one at a time, rather than jumping back over the whole group of steps to the starting point. First, to minimize panic, assure yourself that on the basis of your knowledge of the symptoms and symptom evaluation, you have chosen *all possible* faulty functional units.

For example, you (and the writers of the service literature) may have overlooked a possible faulty unit the first time. The *antenna switch* is common to *both* the transmit and the receive functions. Normally, an antenna switch can produce trouble in either the TRANSMIT or the RECEIVE mode, but not in both. (Typically, an antenna switch is a relay, operated by the TRANSMIT-RECEIVE button or microphone switch. Such relays have been known to stick in one position (TRANSMIT or RECEIVE), but they cannot be in both positions simultaneously.) However, suppose that the relay (antenna switch) is making *poor contact* in both positions. This might result in the symptoms and could make the antenna switch a *possible* (although not likely) faulty unit. Of course, if you cannot logically include other functional units in your original list of selections, you should go back to the symptom-evaluation phase to see if you have overlooked anything.

1-5.8 Verifying Trouble Symptoms

If the equipment design is such that the functional units are plug-in or readily replaceable, the trouble can be verified easily. If you replace a functional unit and the trouble is cleared, this verifies that the replaced unit was at fault (with the possible exception that the plug of the original unit was not making good contact).

When it is not easy to replace a unit, or when no replacement is available, and you must go inside the unit to locate the defective circuit or part, you may save considerable time and effort if you verify the trouble. That is, you should reexamine the possibility that a fault in this unit could logically produce the trouble symptom. You should also ask the question: Does the fault fit the associated information found during symptom evaluation?

For example, if the speech amplifier in the transceiver shown in Fig. 1-10 is defective (produces no output), the transmitted signal will be a constant-amplitude RF signal. There will be no voice information, but the transmitted RF carrier (as well as reception) will be normal. The same symptoms will occur if the modulator is defective. If these are the only symptoms, the choice of either the speech amplifier or the modulator is a good one. Your decision to go inside and make tests on these units as necessary is justified. (This is based on the assumption that you have tried a replacement microphone, as previously discussed.)

1-6. ISOLATING TROUBLE TO A CIRCUIT

The first two steps (symptoms and localization) of the troubleshooting procedure give you the initial symptom information about the trouble and describe the method of localizing the trouble to a probable faulty functioning unit. Both steps involve a minimum of testing except by operation of equipment controls and observation of equipment indicators.

In the isolate step, you will do extensive testing (often with external test equipment) in order to isolate the trouble, first to a group of circuits within the functioning unit and then to the specific faulty circuit. In the case of ICs, (where one IC is used to form a group of circuits within a functional unit), the trouble is isolated to the IC input. No further isolation is necessary because parts within the IC cannot be replaced individually. This same condition is true of some solid-state equipment, in which groups of circuits are mounted on sealed, replaceable boards or cards.

No matter what physical arrangement is used, the isolation process follows the same reasoning you have used previously: the continuous narrowing down of the trouble area by making logical decisions and performing logical tests. Such a process reduces the number of tests that must be performed, thus saving time and decreasing the possibility of error.

1-6.1 Servicing Block Diagrams

Servicing block diagrams provide a pictorial guide for use in isolating the trouble. Although there are variations in formats, the servicing block diagram shown in Fig. 1-19 is typical of those found in military and well-prepared commercial service literature. Figure 1-19 represents the receiver portion of the transceiver shown in Fig. 1-10.

In good service literature, there will be a servicing block diagram for every unit of the equipment, as well as an overall block diagram. In the case of simple equipment, such as those covered by a data sheet only (rather than a complete technical manual), the complete piece of equipment will be represented by one servicing block diagram. (If you are troubleshooting commercial equipment for which service literature is at a bare minimum, there may be no servicing block diagram.)

When using the servicing block diagram to isolate the trouble, you should not discard the symptoms and related information obtained in the previous steps now or at any time during the troubleshooting procedure. From this information, you can identify those circuit groups that are probable trouble sources.

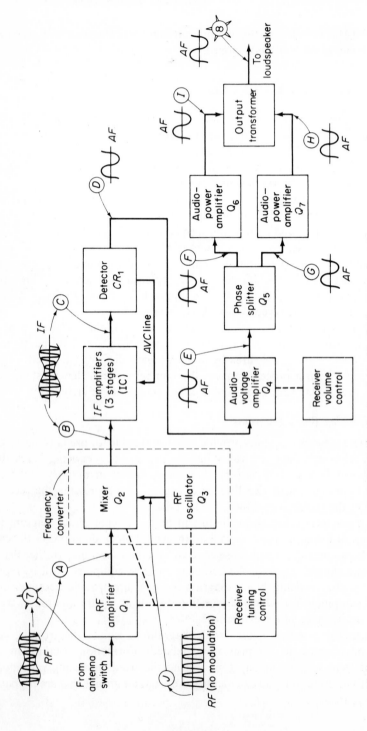

FIGURE 1-19 Servicing block diagram of receiver

Interpreting servicing block diagrams. As the military-type diagram shown in Fig. 1-19 indicates, all circuits that are part of the functional unit are enclosed by a block, as are circuits that make up the circuit groups within the unit. Within each block is the name of the functional unit or circuit group it represents.

Main signal- or data-flow paths are represented by *heavy* solid lines; secondary signal or data paths are represented by *lighter* solid lines. Arrows on the lines indicate the *direction* of signal or data flow.

Operating controls are connected to the circuits that they control by *dotted* or *dashed* lines. When depicted properly, the operating control will be labeled on the diagram; this name should correspond exactly to the control name that appears on the front panel.

Wave forms are given at several points, usually at the input and output of each circuit. Test points are identified by both the *letter* and the *number*. Generally, numbered test points (often starred, as shown here) represent points that are useful in localizing faulty functional units. Lettered test points (often circled, as shown here) represent points that are helpful in isolating faulty circuit groups or individual circuits. (This letter-number combination of test points follows the military style.)

Note that the representation of the circuit groups on the servicing block diagram has no relation to their physical location within the equipment.

Circuit groups versus individual circuits. It is important that you recognize circuit *groups* as well as individual circuits. A *circuit group* is a subdivision of the functional unit that performs a *single electronic subfunction*. (In some cases, a subfunction is performed by an individual circuit.)

If you can subdivide a functional unit into circuit groups rather than individual circuits, you can isolate the group (or remove the group from possible fault) by making a single test at the input or output test point for the group.

For example, there are three circuit groups (frequency converter, IF amplifier, and audio amplifier) and two individual circuits (RF amplifier and detector) in the receiver shown in Fig. 1-19. Each of the three groups and the two individual circuits have input-output test points.

In vacuum-tube equipment, the input signal is injected at the grid, and the output signal is traced at the plate (or possibly at the cathode). In solid-state equipment, the input signal is injected at the base, and the output signal is traced at the collector (or possibly at the emitter). These input-output relationships are shown in Fig. 1-20. For a circuit group, the input is at the *first base* (or grid) in the *signal path*, and the output is at the *last collector* (or plate) in the *same* path. There are certain exceptions to this. For example, in the frequency-converter group shown in Fig. 1-19, the RF oscillator Q_3 has no input but does provide an output to the mixer stage Q_2. Also, the detector stage CR_1 is a diode, with the input at the anode and the output at the cathode.

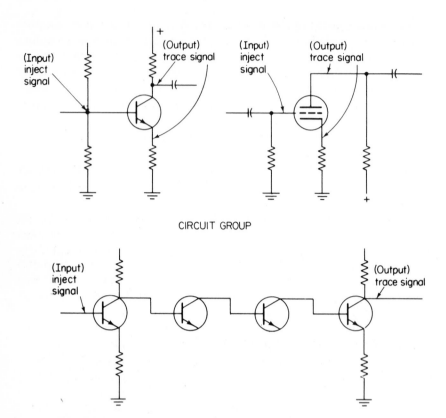

CIRCUIT GROUP

FIGURE 1-20   Input-output relationships in circuit troubleshooting

The input circuit for the complete functional unit shown in Fig. 1-19 is the RF amplifier Q_1, and the base of this transistor is the input signal–injection point. The output from the complete unit is at the secondary winding of the transformer. The input takes place at the point where the functional unit (receiver) receives a signal from another unit (antenna switch); the output supplies a signal to another unit or device (the loudspeaker).

This same idea applies to a circuit group. That is, the input signal to the group is injected at one point, and then the output signal is obtained at a point several stages farther along the signal path. To determine the signal-injection and -output points of a circuit group, you must find the *first* circuit of the group in the signal path and the *final* circuit of the group in the same path. For example, the input for the audio-amplifier circuit group is at the base of Q_4 (test point D); whereas the output is at the secondary winding of the transformer (test points H or I, or possibly 8). Note that test point D is simultaneously at the input of the audio amplifier circuit group and the output of the detector circuit.

Note that the IF amplifier circuit group is an integrated circuit consisting of three amplifier stages. Although each stage has an input and an output, none of the stages has separate input-output test points; whereas the various stages in the audio amplifier circuit group (which is made up of separate, replaceable parts) do have separate test points. Even if it were possible, there would be no value in testing each stage of the IF amplifier because the entire IC must be replaced as a package.

1-6.2 Signal Paths

There are six basic types of signal paths, no matter which circuit group or circuit arrangement is used. There are shown in Fig. 1-21 and are summarized in the following paragraphs:

A *linear path* is a series of circuits arranged so that the output of one circuit feeds the input of the following circuit. Thus, the signal proceeds straight through the circuit group without any return or branch paths. The

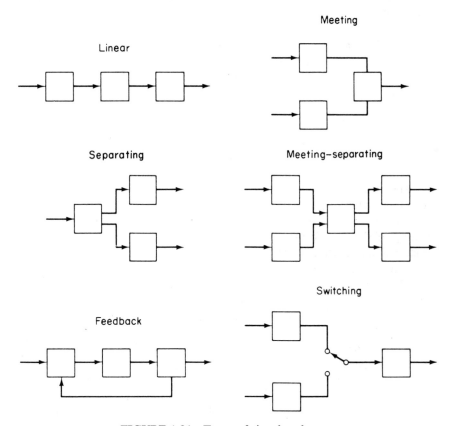

FIGURE 1-21   Types of signal paths

main signal path through the IF amplifiers shown in Fig. 1-19 is an example of a linear path.

A *meeting path* is one in which two or more signal paths enter a circuit. The paths to the mixer Q_2 and the output transformer shown in Fig. 1-19 are both meeting paths.

A *separating path* is one in which two or more signal paths leave a circuit. In Fig. 1-19, the paths from the phase splitter Q_5 are separating paths.

A *meeting-separating path* is one in which a single stage has multiple inputs and multiple outputs.

A *feedback path* is a signal path from one circuit to a point or circuit preceding it in the signal-flow sequence. In Fig. 1-19, the automatic volume control (AVC) line from the detector to the IF amplifiers is an example of a feedback path.

A *switching path* contains a selector switch (or a similar device, such as a relay) that provides a different signal path for each switch position. In Fig. 1-10, the paths from the transmitter and receiver to the antenna assembly (through the antenna switch) are switching paths.

1-6.3 Signal Tracing versus Signal Substitution

Both signal tracing and signal substitution (or signal injection) are used frequently in troubleshooting all types of electronic equipment, including digital equipment.

Signal tracing is done by examining the signal at a test point with a monitoring device such as an oscilloscope, a multimeter, or a loudspeaker. A signal is applied at a fixed point, and the input probe of the indicating or monitoring device is moved from point to point. The applied signal can be generated from an external device, or the normal signal associated with the equipment can be used (e.g., using the regular broadcast signal to trace through a radio receiver).

Signal substitution is done by injecting an artificial signal (from a signal generator, a sweep generator, or a similar device) into a circuit or a complete functional unit to check performance. The injected signal is moved from point to point, and the indicating or monitoring device remains fixed at one point. The monitoring can be done with external test equipment or with the monitoring device associated with the equipment (e.g., using the loudspeaker of a radio receiver to monitor signals injected at various points).

Signal tracing and signal substitution are often used simultaneously in troubleshooting. For example, when troubleshooting a solid-state audio system, an audio generator can be used to inject a signal at the input while an oscilloscope can be used to observe the wave form at each stage or circuit. Or a *repetitive pulse* can be introduced into a digital system by means of a pulse generator. The resultant wave forms at various gates, flip-flops, counters, and so forth are then observed on an oscilloscope.

1-6.4 Half-Split Technique

The *half-split technique* is based on the idea of using any test to eliminate the maximum number of circuit groups or circuits simultaneously. This will save both time and effort. The half-split technique is used primarily in isolating trouble in a linear signal path, but it can be used with other types of signal paths. In this system, brackets are placed at good-input and bad-output points in the normal manner, and the symptoms are studied. Unless the symptoms point *definitely* to one circuit group or circuit that might be the trouble source, the most logical place to make the first test is at a *convenient* test point *halfway between* the brackets.

Example of the half-split technique. The block diagram shown in Fig. 1-22 is a simplified version of the servicing block diagram for a receiver shown in Fig. 1-19. Figure 1-22 shows the linear signal path of the received signal through the receiver unit by showing the circuit groups consolidated into *single* blocks. The brackets placed at test point 7 (good input) and test point 8 (bad output) show the trouble being localized to the receiver unit.

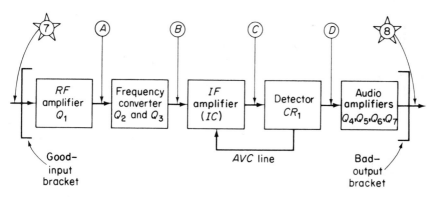

FIGURE 1-22 Simplified servicing block diagram of receiver unit illustrating the linear signal path of received signal through the circuit groups

The next phase of troubleshooting is to isolate the trouble to one of the circuit groups (frequency converter, IF amplifier, or audio amplifier) or to one of the individual circuits (RF amplifier or detector) in the linear signal path. The selection of test points during this phase depends upon their *accessibility* and the *method* of troubleshooting (signal tracing or signal injection).

Assuming that test points *A*, *B*, *C*, and *D* are equally accessible (and that there are no special symptoms that would point to a particular circuit or group), test point *C* is the most logical point for the first test if signal tracing is used. Test point *B* is the next logical choice. With signal injection, however, test points C and D are the most logical choices. The discussion that follows describes the reasoning for making these choices.

Half-split technique using signal tracing. If signal tracing is used, an RF signal (at the receiver frequency) modulated by an AF tone is introduced at test point 7. An oscilloscope is then connected to monitor the wave form at various test points, as shown in Fig. 1-23.

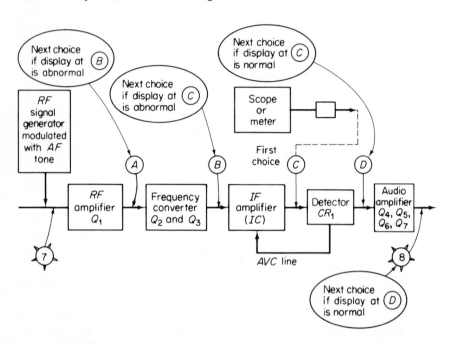

FIGURE 1-23   Example of half-split technique using signal tracing

Test point C is the most logical choice for a first test. If C is chosen first and the oscilloscope display is normal, you have cleared four circuits (RF amplifier, mixer, RF oscillator, and IC IF amplifier) but there will still be circuits that are possibly defective. This process divides the circuits into two groups (known good and possibly bad).

Now, assume that the indication at test point C is abnormal. The bad-output bracket can be moved to test point C, with the good-input bracket remaining at test point 7. The next logical test point is B, because it is near the *halfway point* between the two brackets. (If you choose point A instead of point B, you will confirm or deny the possibility of trouble in the RF amplifier *only*.)

If the oscilloscope display is abnormal at test point B, the trouble is isolated to the RF amplifier or the frequency converter. Additional observation at test point A will further isolate the trouble to either circuit.

The final step in this troubleshooting process is to monitor the signal at test point A. If there is an abnormal indication at A, then the bad-output bracket can be moved to A and the trouble isolated to the RF amplifier. If there is a normal display on the oscilloscope at A, the good-input bracket can be moved to A and the trouble isolated to the frequency converter.

Now, let us see what happens if a test point *other than C* is monitored first, using signal tracing.

If you choose test point A for the first test and the oscilloscope display is normal, the trouble is located somewhere between A and the receiver output (test point 8). This means that the trouble could be in the frequency converter, IF amplifier, detector, or audio amplifier. You still have many test points to check.

On the other hand, if an abnormal signal is observed at test point A, the trouble is immediately isolated to the RF amplifier. All other circuits are eliminated, but this would be a *lucky accident*, not good troubleshooting. To be performed efficiently and rapidly, the troubleshooting procedures should be based on a systematic, logical process, not on chance or luck. You will probably have as many unlucky accidents as lucky ones throughout your troubleshooting career.

The same condition is true if test point D is chosen first (if you use signal tracing). A normal oscilloscope display will clear all circuits except the audio section. However, an abnormal display will still leave the possibility of trouble in many circuits.

If test point B is chosen first and the oscilloscope display is normal, this will clear three circuits (RF amplifier, mixer, RF oscillator) but will leave six circuits possibly defective (the IC IF amplifier, detector, audio-voltage amplifier, phase splitter, and both audio-power amplifiers). The opposite results will be obtained if the oscilloscope display is abnormal.

Actually, test point B is not a bad choice for a first test using signal tracing. If test point B is more readily accessible than test point C, then use it. For example, many television receivers have a test point B (often called the *looker* point) on the outside of the RF tuner shield. This is sometimes easier to reach than test point C, which is at the input to the detector and may be located on the underside of the television chassis.

Half-split technique using signal injection. If signal injection is used, signals of the right sort are injected at test points A, B, C, D, and 7, as shown in Fig. 1-24. Receiver response is noted on the loudspeaker (test point 8). In this case, the signal for test points A and 7 is at the RF frequency of the receiver, modulated by an AF tone. The signals for B and C are at the IF frequency, also modulated by an AF tone. The signal for D is an AF tone.

Note that signal injection requires several different types of signal sources at different frequencies, whereas signal tracing requires only one signal at the input. (In some cases, the input can be the signal that is normally

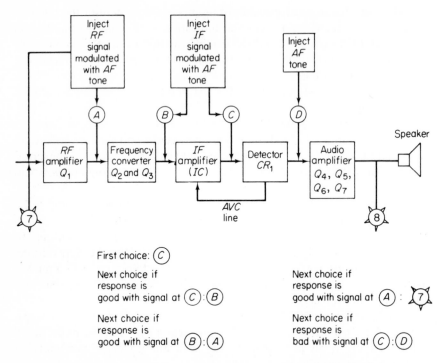

FIGURE 1-24   Example of half-split technique using signal injection

associated with the equipment. For example, you can use a broadcast signal as the signal source when troubleshooting a radio receiver. Of course, this advantage does not always exist in all equipments.)

Using signal injection with the half-split technique, the first signal is again injected at test point *C*. Now, however, a normal response from the loudspeaker (with a signal at *C*) will clear the final five circuits (detector and four audio circuits). Under these circumstances, the next logical points for signal injection are *B* and *A*, in that order.

An absent or abnormal response (with a signal injected at *C*) will isolate the trouble to the detector and audio circuits. Under these circumstances, the next logical test point is *D*.

A normal response from the loudspeaker with a signal at *D* but not with a signal at *C* will isolate the trouble to the detector. An absent or abnormal response with a signal at *D* isolates the trouble to the audio circuits.

Keep in mind that these examples using the half-split technique, signal tracing, signal injection, and bracketing to isolate the trouble to a circuit group by no means cover all the possibilities that may occur. They simply illustrate the basic concepts involved when following the systematic, logical troubleshooting procedure.

1-6.5 Isolating Trouble to a Circuit within a Circuit Group

Once trouble is isolated to a faulty circuit group in the receiver unit, the next step is to isolate the trouble to the faulty circuit within the group. Bracketing, half-splitting signal tracing, signal injection, and *knowledge of the signal path* in the circuit group are all important to this step and are used in essentially the same way they were used for isolating trouble to a circuit group.

Isolating the trouble to a circuit in a linear path is identical to the process described thus far. However, isolating trouble to a circuit in a *separating* or *meeting* path requires a variation in the procedure; this is also true for isolating trouble in *feedback* and *switching* paths. Although these various signal paths are not recognized so easily as linear paths, you should have little difficulty in recognizing them if you follow the definitions given in Sec. 1-6.2.

Sometimes, various types of signal paths are combined. For example, the audio-circuit group of the receiver shown in Fig. 1-25, contains a linear path (from audio-voltage amplifier to phase splitter), a separating path (from phase splitter to audio-power amplifiers), and a meeting path (from audio-power amplifiers to output transformer).

Example of isolating trouble to a circuit (separating and meeting paths). In Fig. 1-25, the audio amplifier circuit group is illustrated in a manner slightly different from that used in the servicing block diagram of Fig. 1-19. The good-input bracket at test point D and the bad-output bracket at test point 8 indicate that the trouble is isolated to the circuit group. There is a normal signal (injected from an audio generator) at the input base of the audio-

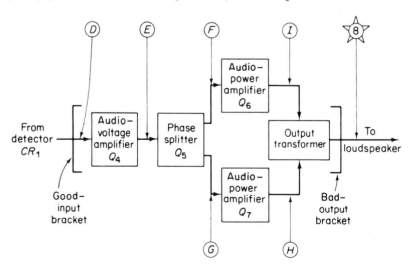

FIGURE 1-25   Receiver audio section showing separating and meeting paths

voltage amplifier Q_4, and no signal is observed (on an oscilloscope) at the output transformer.

Before you can isolate the trouble in separating or meeting signal paths, you must isolate these paths from the rest of the circuit group. That is, the phase splitter, audio-power amplifiers, and output transformer must first be isolated from the audio-voltage amplifier. This is done by making the first test at point E.

Assume that the oscilloscope is now moved to test point E, as shown in Fig. 1-26, and that a normal signal is observed there. This means that the audio-voltage amplifier Q_4 is functioning normally. The good-input bracket can be moved to test point E, and the trouble is isolated to the phase splitter, power amplifiers, or output transformer. (If the signal is abnormal at E, the trouble is isolated immediately to the audio-voltage amplifier.) With a normal signal at E, the trouble is isolated to the separating or meeting signal paths in the circuit group.

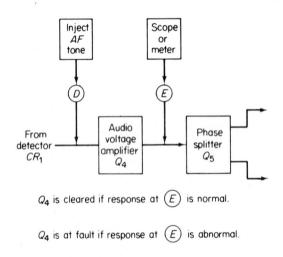

Q_4 is cleared if response at $\text{\textcircled{E}}$ is normal.

Q_4 is at fault if response at $\text{\textcircled{E}}$ is abnormal.

FIGURE 1-26 First step in isolating trouble in separating and meeting paths

The next step is to eliminate one of the paths and thus isolate the trouble to the other path. For troubleshooting purposes, the remaining circuits can be considered to be two *linear* paths, as shown in Figs. 1-27 and 1-28. One path (Fig. 1-27), includes the phase splitter Q_5, power amplifier Q_6, and the transformer. The other path (Fig. 1-28), includes the phase splitter Q_5, power amplifier Q_7, and the transformer. Isolating trouble to these paths involves checking signals at test points F, G, H, and I.

To check the top path (Fig. 1-27), an audio signal is injected at E, and an oscilloscope is used to monitor test points I and F (in that order). To check

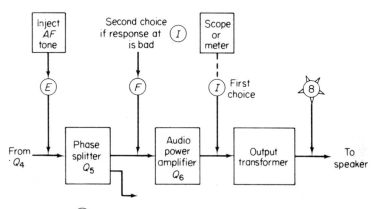

If response at I is good, check Q_7 half of circuit.

FIGURE 1-27 Checking Q_6 half of separating-meeting signal path

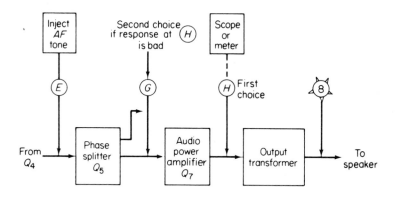

If response at H is good, check Q_6 half of circuit.

FIGURE 1-28 Checking Q_7 half of separating-meeting signal path

the bottom path (Fig. 1-28), the audio signal is still injected at E, and the oscilloscope is used to monitor H and G (in that order). Note that if the first tests are made at F or G, nothing is proved if the signal is normal. Thus, the first tests should be made at H or I.

If the signal is normal at I, the top signal path is eliminated, and the next check should be made at H. If the signal is abnormal at H, the bad-output bracket is moved to H. Then, test point G is the next logical point for a check. If the indication at G is normal, the good-input bracket is moved to G, and the trouble is isolated to audio-power amplifier Q_7. If the indication at G

is abnormal, the bad-output bracket is moved to G, and the trouble is isolated to phase splitter Q_5.

If the signal is abnormal at I, trouble is isolated to the top signal path; test point F is the logical point for a check. If the indication at F is normal, the good-input bracket is moved to F, and the trouble is isolated to audio-power amplifier Q_6. If the indication at F is abnormal, the bad-output bracket is moved to F, and the trouble is isolated to phase splitter Q_5.

The thought process involved for isolating trouble in the audio-circuit group is illustrated in the *flow diagrams* shown in Figs. 1-29 and 1-30. The flow diagrams are similar to those used in the troubleshooting sections of some service literature, particularly military technical manuals. The question "Is signal normal?" means "Is the signal response *at that test point* normal?" Note that Fig. 1-29 is based on the assumption that the initial test is made at test point I (in the top signal path) after the test at E. Figure 1-30 is based on an initial test at H (in the bottom signal path) after the test at E.

Example of isolating trouble to a circuit (feedback path). As discussed, a feedback path is a signal path from one circuit to a point or circuit preceding

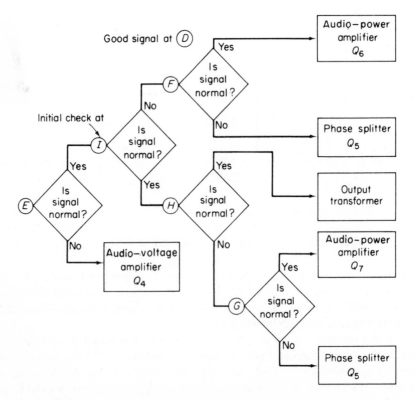

FIGURE 1-29 Thought process for isolating trouble in audio circuit group (initial test made at I, after E)

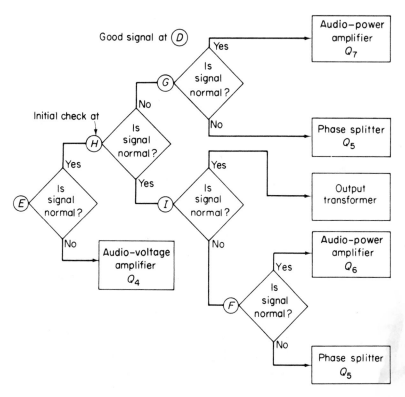

FIGURE 1-30 Thought process for isolating trouble in audio circuit group (initial test made at *H*, after test at *E*)

it in the signal-flow sequence. In the receiver under consideration, a feedback path is represented by the AVC line from the detector CR_1 to the bases of the IF amplifiers in the IC module, as illustrated in Fig. 1-31.

There are two outputs from the circuit of Fig. 1-31. The main signal output is the audio signal from detector CR_1 to the voltage amplifier Q_4. This output is monitored at test point *D*. The other output is the AVC feedback voltage from the detector to the IF amplifiers. When a strong signal is being passed through the IF stages, the AVC output voltage of the detector prevents the audio (main signal flow) from becoming too strong. A weak IF signal, however, does not result in a high AVC voltage. Thus, there is enough gain in the circuit to amplify the weak signal.

Test connections. In troubleshooting the circuit shown in Fig. 1-31, an IF signal (modulated by an AF tone) is injected at test point *B*. Test points *C* and *D* are monitored with an oscilloscope. The AVC output appears as a dc voltage and is monitored with a voltmeter. Note that the AVC line is not assigned a test point number or letter. However, there is a note indicating that the AVC voltage can be measured across R_7 (a resistor in the detector circuit).

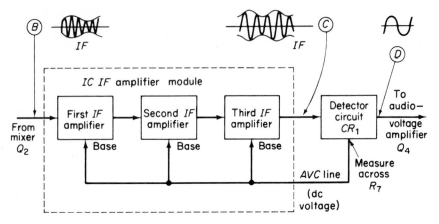

FIGURE 1-31   Example of feedback path (AVC line)

A similar note or test-point assignment will be found in service literature. In well-prepared literature, the physical location of test points and important circuit parts is given by means of photographs or drawings. Unfortunately, in some service literature, you must find the *electrical* location on the schematic diagram and then hunt for the *physical* location, possibly with the aid of an illustrated list of parts.

Troubleshooting sequence. If the trouble is isolated to the circuits shown in Fig. 1-31, first examine both detector outputs and find out if the fault is most likely in the main signal path (test point *D*) or in the feedback path (voltage across R_7), as shown in Fig. 1-32.

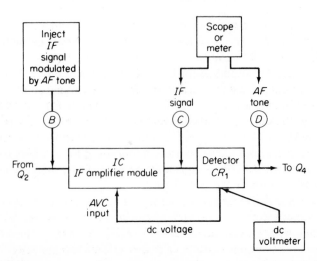

FIGURE 1-32   Troubleshooting connections for isolating trouble in feedback path

If both detector outputs are weak, or if there is no output whatsoever with an input applied to the IF module (at test point B), the half-split method can be applied to the main signal path. This is done by monitoring the IF signal at test point C.

On the other hand, an output signal at test point D that is considerably stronger than normal for a given input is an indication that the fault is in the feedback path. In this case, the AVC voltage output of the detector should be checked.

A low AVC voltage indicates that the trouble is in the detector circuit. In vacuum-tube receivers, the AVC voltage is generally negative because it is applied to tube grids (which require a negative to reduce gain). In solid-state equipment, the AVC feedback voltage can be positive or negative, depending upon whether the IF amplifier transistors are positive-negative-positive or negative-positive-negative. Either way, the AVC output must provide an *opposing* voltage that has a value high enough to decrease the gain of the IF amplifiers when a strong signal is received.

If the opposing voltage is low, or if there is no AVC voltage whatsoever, there will be no feedback signal. Consequently, a strong signal at the input of the IF amplifiers will be amplified at a high level, causing a stronger-than-normal signal at the audio output of the detector. The thought process involved for isolating trouble in the AVC feedback circuit is illustrated in the flow diagram shown in Fig. 1-33 (which is based on the assumption that the initial test is made in the main audio-signal path at test point D).

Alternate methods for feedback circuits. Another method (in addition to measuring the voltage) of troubleshooting circuits that contain feedback paths is to *disable* the feedback loop. This may be done by disconnecting the feedback path (opening the loop) or by shorting the feedback-signal path to ground. Opening the loop is sometimes inconvenient, and shorting the AVC signal to ground could cause damage to the circuit. *Never short any circuit to ground unless you are certain that no damage will result.*

One method for disabling a feedback line without damage is to lift the line at the source and connect a resistance between the source and ground, as shown in Fig. 1-34. In this case, you disconnect the line from the detector, as shown in Fig. 1-34(a); measure the resistance of the line (to the IF module), as shown in Fig. 1-34(b); and then connect an equivalent resistance between the source (detector circuit) and ground, as shown in Fig. 1-34(c). In this way, you can disable the AVC line and measure the available AVC voltage across the equivalent resistance.

Another method for disabling the feedback line is to apply a fixed bias to the line, as shown in Fig. 1-35. The fixed bias must be large enough to override any feedback voltage. For example, if the AVC feedback voltage is about 1 or 2 V, the fixed bias must be about 3 or 4 V. This method of disabling the AVC line is often used in troubleshooting television receivers, as discussed in later chapters.

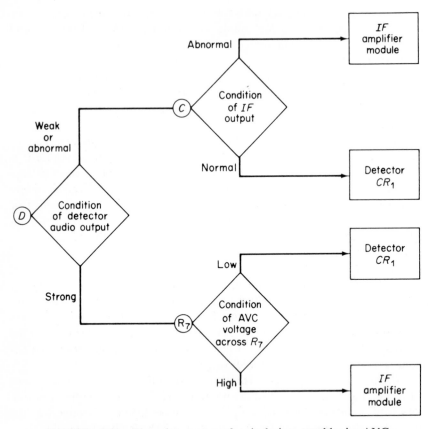

FIGURE 1-33 Thought process for isolating trouble in AVC feedback circuit (initial test made at D)

Note that some receivers (particulary communications-type receivers) are provided with an ON-OFF switch for the AVC line. If trouble is located in the circuit containing the AVC loop, the switch can be used to find out if the trouble still exists without AVC feedback voltage.

Example of isolating trouble to a circuit (switching path). In order to isolate the faulty circuit in a switching path, you must first test the final output for the circuit *following* the switch at both switch positions. When the switch is a multiple-contact type, each contact may be connected to a different circuit. In this case, it may be necessary to place the switch in each position and check the final output of the circuit associated with that position. If the symptoms and/or tests point to one specific circuit, it may not be necessary to check every switch position.

Once this test has been performed and the trouble isolated to one or more of the branches, the suspected branches are checked to locate the one that is faulty. The next step is to apply the techniques of half-splitting and isolating in

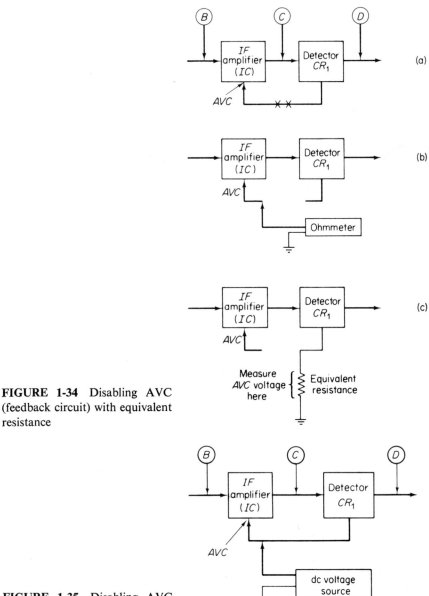

FIGURE 1-34 Disabling AVC
(feedback circuit) with equivalent
resistance

FIGURE 1-35 Disabling AVC
(feedback circuit) with bias
voltage

the usual manner. Because a switch connects different combinations of circuit
groups, it can be used to help isolate the trouble to one or more of the circuits.

For example, consider the switching path represented in Fig. 1-36,
which shows the IF circuits of a receiver. There are two separate IF signal

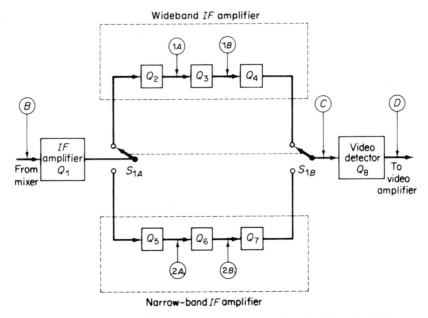

FIGURE 1-36 Example of switching circuit signal paths (IF circuits of radar receiver)

paths (wide band and narrow band), that can be selected by switch S_1. The two sections of S_1 are ganged together.

By observing the output of video detector Q_8 (test point D) with a proper IF signal injected at the input of the first IF amplifier Q_1 (test point B), you can use switch S_1 to help isolate the trouble.

If the output of Q_8 is incorrect only during wideband operation, the trouble is in the wideband circuit group Q_2, Q_3, and Q_4. If the output of Q_8 is incorrect only during narrow-band operation, the trouble is in the narrow-band circuit group Q_5, Q_6, and Q_7. If the abnormal symptoms are present for both positions of the switch, the trouble is in the first IF amplifier Q_1 or in the video detector Q_8 (or in the switch S_1). The rest of the troubleshooting procedure consists of narrowing down the trouble to a single circuit within the faulty circuit group.

For example, assume that a check at test point D shows an abnormal signal for both positions of S_1 when a normal signal is injected at test point B. The next logical place for signal tracing is test point C.

Because both the narrow-band and wideband modes of operation are abnormal at the video-detector output, the trouble must be in a circuit common to both modes. In this case, the trouble is in either the video detector Q_8 or the first IF amplifier Q_1. Checking for a signal at test point C will isolate the trouble to one of these two circuits. All the circuits in the narrow-band

and wide-band circuit groups (Q_2 to Q_7) are eliminated because a signal in any one of these circuits will produce an abnormal signal only for its respective mode of operation.

With a normal signal at C, the trouble is isolated to the video detector Q_8. An abnormal signal at C (in both positions of the switch) establishes the trouble at the first IF amplifier Q_1.

The thought process involved for isolating trouble in the switching circuit is illustrated in the flow diagram shown in Fig. 1-37. This process is based on the assumption that the *initial* test is made at test point D with a good-input signal at B.

For conditions in which the trouble is isolated to the narrow-band or the wide-band circuit groups, the flow diagram (Fig. 1-37) illustrates the respective first tests being made at the output of the second IF amplifier, Q_3 or Q_6,

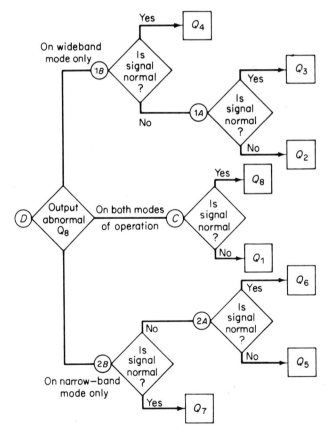

FIGURE 1-37 Thought process for isolating trouble in switching path (good signal at B, initial test made at D)

depending on the mode of operation. The flow diagram will be slightly different if the first test is made at the output of the first amplifier in each case. However, the thought process and the logic are identical.

The test sequence illustrated in Fig. 1-37 does not include testing for a faulty switch S_1. If trouble is isolated to a circuit that is connected to one of the switch contacts, a check should be made between the circuit and the switch before the circuit is bracketed.

1-7. LOCATING A SPECIFIC TROUBLE

The ability to recognize symptoms and to verify them with test equipment will help you to make logical decisions regarding the selection and localization of the faulty functional unit. This ability will also help you to isolate trouble to a faulty circuit. The final step of troubleshooting, locating the *specific* trouble, requires testing of the various branches of the faulty circuit to find the defective component.

The proper performance of the locate step will enable you to find the cause of trouble, repair it, and return the equipment to normal operation. You should follow up this step by making a written record of the trouble so that, on the basis of this history of the equipment, future troubles may be easier to locate. Also, such a history may point out consistent failures that could be caused by a design error.

1-7.1 Locating Troubles in Plug-in Modules

Because so much modern electronic equipment is of IC and sealed-module design, technicians often assume that it is unnecessary to locate specific troubles to individual components. That is, they assume that *all* troubles can be repaired by replacement of sealed modules. Some technicians are even trained that way. But the assumption is not true.

Although the use of replaceable modules often minimizes the number of steps required in troubleshooting, it is still necessary to check circuit branches to components outside the module. Front-panel operating controls are a good example of this because such controls are not located in the sealed units. Instead, they are connected to the terminal of an IC, circuit board, or plug-in module.

1-7.2 Inspection Using the Senses

After the trouble is isolated to a circuit, the first step in locating the trouble is to perform a preliminary inspection using the senses (sight, smell, hearing, and touch). For example, burned or charred resistors can often be spotted by visual observation or by smell. The same holds true for oil-filled or wax-filled components such as some capacitors, inductors, and transformers.

When overheated, the oil or wax expands and can leak out or perhaps cause the case to buckle or even explode. Overheated components, such as hot transistor cases, can be located quickly by touch. The sense of hearing can be used to listen for high-voltage arcing between wires or between wires and the chassis, for cooking of overloaded or overheated transformers, or for hum or lack of hum, whichever the case may be.

Although the senses of sight, smell, hearing, and touch are used at this time, the procedure is referred to more frequently as a *visual inspection*. It is possible to find a defective component by visual inspection, but the component should not be replaced until the trouble has been investigated further. Often, the real cause of the trouble will destroy another component. For example, a short can cause a resistor to burn out. If the resistor is replaced without first repairing the short, the new resistor will also burn out. Therefore, always check for *other possible troubles* after locating a defective component.

1-7.3 Testing to Locate a Faulty Component

Testing vacuum tubes. Vacuum tubes are relatively easy to replace (compared with transistors or IC units). For this reason, two common practices for troubleshooting vacuum-tube equipment have been developed over the years. One practice is to test all vacuum tubes by substitution as a first step in troubleshooting. The other is to remove and test all tubes on a tube tester as a first troubleshooting step. Neither practice is valid.

With *tube substitution*, the usual procedure is to replace the tubes one at a time until the equipment again works normally. The *last tube* replaced is then discarded, and all the other original tubes are reinserted in their respective sockets.

There are several fallacies in this practice. Some oscillator circuits or high-frequency circuits may operate with one new tube but not with another because of the differences in interelectrode capacitance between the tube elements (a good tube may react like a bad tube). If you rock or rotate the tubes while removing or inserting them, the result may be bent pins or broken weld wires where the pins enter the tube envelope. If there is more than one bad tube in the equipment at any time, substituting good tubes one at a time and reinserting the original tube before substituting the next tube will not locate the defective tubes. Finally, if the replacement tube becomes defective immediately after substitution, there definitely is circuit trouble, and further troubleshooting is required anyway.

Testing all tubes on a tube tester as a first troubleshooting step is also *not* recommended. Because it has been followed religiously in the past, this practice has led to the misconception that defective tubes are the cause of the majority of equipment failures. Studies have shown that when tubes removed from equipment have been checked further, many have proved not defective. This leads to the conclusion that if all tubes are left in their sockets, tubes will probably prove to be the cause of considerably less than 50 percent of equipment troubles. Even if the percentage is higher, the process of removing tubes, checking them on a tube tester that may be marginal, and replacing them with new tubes is a waste of time, as well as a poor troubleshooting practice.

It is not to be inferred that the tubes should never be checked first once the trouble is isolated to the circuit. For example, the power can be turned on and the tube filaments checked first for *proper warm-up*. If the tube envelopes are glass, a visual inspection will show whether the filament is burned out. If the tube envelopes are metal, you can feel the envelope to find out whether the filament is lit.

This type of test will often speed the troubleshooting effort by quickly locating a tube having a burned-out filament. If a tube does not warm up properly, remove it and check it on a tube tester or substitute a new (known-good) tube, whichever is most convenient. In either case, the complete circuit should be checked to determine whether the tube burned out naturally from long use or from some trouble in the circuit. Simply replacing the tube without checking the rest of the circuit does not complete the location of trouble. You still must verify whether the burned-out tube is the cause or the effect of the trouble.

The preceding paragraph described the procedure for checking the tube when the tube filaments are all connected *in parallel*. In that case, when the filament of one tube burns out, that tube (and only that tube) will show a bad (unlit) filament. However, the filaments of tubes connected *in series* present more of a problem. When one filament burns out, all the filaments in the string will be unlit. This makes it more difficult to determine which filament in the string is the bad one. (See Fig. 1-38.)

Removing the tubes one at a time and checking their filaments for continuity with an ohmmeter is time consuming, and if care is not taken, the

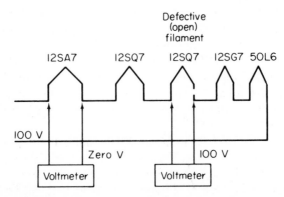

FIGURE 1-38 Method of locating defective (open) filament in vacuum-tube equipment (with filaments connected in series)

tube may be damaged during the test because the current from the ohmmeter (set on its lowest scale) may be high enough to burn out the filament.

A better test is to measure the voltage *across* the filament terminals of the tube socket (if the bottom of the socket is accessible and all the tubes are left in their sockets). All good filaments in the string will show zero voltage, but the one that is defective (burned out) will have the full voltage that is applied to the filament string, as shown in Fig. 1-38.

Testing solid-state and IC equipment. Unlike vacuum tubes, transistors, ICs, and solid-state diodes are not easily replaced. Thus, the old electronic troubleshooting procedure of replacing tubes at the first sign of trouble has not been carried over into the repair of solid-state equipment. Instead, solid-state circuits are analyzed by *testing* in order to locate faulty components.

Testing the active device. For service and maintenance purposes, the vacuum tube, transistor, IC, and solid-state diode may be considered the active devices (or common denominators) in an electronic circuit. Because of their key position in the circuit, these devices are a convenient point for evaluating the operation of the entire circuit through wave-form, voltage, and resistance tests. Making these tests at the terminals of the active device will usually result in locating the trouble quickly.

Wave form testing. Usually, the first step in circuit testing is to analyze the output wave form of the circuit, or the output wave form of the active device (generally the plate of a vacuum tube or the collector of a transistor). Of course, in some circuits (such as power supplies), there is no output wave form. And in other circuits, there is no wave form of any significance.

In addition to being displayed on an oscilloscope, the wave forms must be analyzed in detail to check *amplitude, duration, phase,* and/or *shape.* As is discussed in Chapters 2 and 3, a careful analysis of wave forms can pinpoint the branch of a circuit that is most likely to be defective.

Transistor and diode testers. It is possible to test transistors and diodes in circuit by using in-circuit testers. (Such testers are discussed further in Chapter 2.) These testers are usually quite good for transistors used at lower frequencies, particularly in the audio range. However, most in-circuit testers will not show the *high-frequency* and/or *switching* characteristics of transistors. (The same is true for out-of-circuit transistor and diode testers.) For example, it is quite possible for a transistor to perform well as an audio amplifier but to be hopelessly inadequate as a high-speed switching device required in digital equipment.

Voltage testing. After wave-form analysis and/or in-circuit tests, the next logical step is voltage measurement at the terminals or leads of the active device. Always pay particular attention to those terminals that show an abnormal wave form. These are the terminals most likely to show an abnormal voltage.

When properly prepared service literature is available (with wave-form, voltage, and resistance charts), the *actual* voltage measurements can be compared with the *normal* voltages listed in the voltage charts. This test will often help to isolate the trouble to a single branch of the circuit.

Relative voltages in solid-state equipment. Often you will have to troubleshoot solid-state circuits without benefit of adequate voltage and resistance information. This can be done by using the schematic diagram to make a logical analysis of the relative voltages at the transistor terminals. For example, with an *npn* transistor, the base must be positive in relation to the emitter if there is to be emitter-collector current flow. That is, the emitter-base junction will be forward-biased when the base is more positive (or less negative) than the emitter. The problem of troubleshooting solid-state circuits on the basis of relative voltages is discussed further in Chapter 3.

Resistance measurements. After you have made wave-form and voltage measurements, it is often helpful to make resistance measurements at the same point on the active device (or at other points in the circuit), particularly where an abnormal wave form and/or voltage is found. Suspected parts can often be checked by a resistance measurement, or a continuity check can be made to find the point-to-point resistance of the suspected branch. As we have noted, considerable care must be used when making resistance measurements in solid-state circuits. The junctions of a transistor act like diodes. When properly biased (by the ohmmeter battery), the diodes will conduct and produce false resistance readings. This condition is discussed in Chapters 2 and 3.

Current measurements. In rare cases, the current in a particular circuit branch must be measured directly with an ammeter. However, it is usually simpler and more practical to measure the voltage and resistance of a circuit and then calculate the current.

1-7.4 Wave-form Measurements

When you are testing to locate trouble, wave-form measurements are made with the circuit in operation and usually with an input signal or signals applied. If you use the wave-form reproductions found in some service literature, follow all the notes and precautions described. The literature will usually specify the position of operating controls, typical input voltage, and so on.

Figure 1-39 shows wave-form reproductions found in military-type service literature. Such reproductions are actual photographs of wave forms taken with an oscilloscope from equipment that is operating properly. Note that the scale divisions of the oscilloscope screen are given in these reproductions. The horizontal scale permits the wave-form duration, period, frequency, and phase relationship to be measured. The vertical scale is used to measure wave-form amplitude (voltage). The use of an oscilloscope to duplicate such measurements is discussed in Chapter 2. The type of wave-form reproductions shown in Fig. 1-39 are most useful in troubleshooting

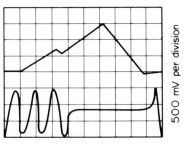

FIGURE 1-39 Example of wave forms found in military-type service literature

situations in which information about frequency and phase are critical to proper circuit operation.

Figure 1-40 shows wave-form reproductions found in typical commercial service literature. That is, the approximate shape of the wave forms is

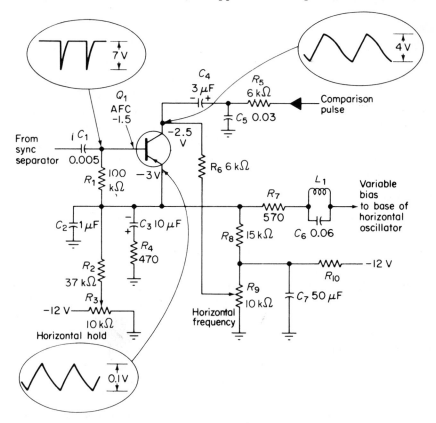

FIGURE 1-40 Examples of wave-form and voltage information found in commercial service literature

given on the schematic diagram, together with the voltage amplitude. There may or may not be a note on the diagram indicating the approximate frequency of the wave forms.

It is obvious that the wave forms shown in Fig. 1-39 are far superior to those shown in Fig. 1-40. However, keep in mind that there is a relationship between wave forms and trouble symptoms. Complete failure of a circuit will usually result in the absence of a wave form. A poorly performing circuit will usually produce an abnormal or distorted wave form. Furthermore, exact wave forms are not always critical in all circuits.

1-7.5 Voltage Measurements

When you are testing to locate trouble, the voltage measurements are made with the circuit in operation but usually with no signals applied. If you are using the voltage information found in the service literature, follow all the notes and precautions on the charts. The charts will usually specify the position of operating controls, typical input voltages, and so on.

In commerical service literature, the voltages are often shown on the schematic diagram, along with the wave forms, as illustrated in Fig. 1-40. This system is quite accurate, but it does require that you find the actual physical location of the terminals where the voltages are to be measured.

Figure 1-41 shows voltage information as it often appears in military-type service literature. In Fig. 1-41(a), the voltage information is simply listed in chart form. In Fig. 1-41(b), the physical arrangement of the terminals is given along with the normal voltage to be measured at each terminal. In many cases, the physical relationship of the tubes (or other active devices) is given, as shown in Fig. 1-41(c).

Because of the safety practice of setting a voltmeter to its highest scale before making measurements (see Chapter 2), the terminals having the highest voltage (vacuum-tube *plate* or transistor *collector*) should be checked first. (In some solid-state equipment, the collector is grounded, and the emitter has the highest voltage.) Then, the elements having lesser voltage should be checked in descending order.

If you have had any practical troubleshooting experience, you know that voltage (as well as resistance and wave-form) measurements are seldom identical to those listed in the service literature. (The same is true of troubles listed in service literature trouble charts. In actual maintenance, you will rarely find trouble that appears in the manual.)

This brings up an important question concerning voltage measurements: How close is good enough? There are several factors to consider in answering this question.

The tolerances of the resistors, which greatly affect the voltage readings in a circuit, may be 20, 10, or 5 percent. Resistors with 1 percent tolerance (or better) are used in some critical circuits. For this reason, the tolerances

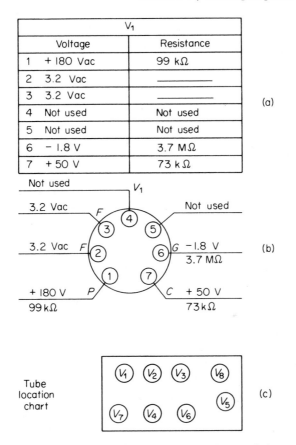

V₁	
Voltage	Resistance
1 + 180 Vac	99 kΩ
2 3.2 Vac	————
3 3.2 Vac	————
4 Not used	Not used
5 Not used	Not used
6 − 1.8 V	3.7 MΩ
7 + 50 V	73 k Ω

(a)

(b)

Tube
location
chart

(c)

FIGURE 1-41 Examples of voltage and resistance information found in military-type service literature

marked or color-coded on the parts are an important factor. Transistors and diodes have a fairly wide range of characteristics and thus will cause variations in voltage readings.

The accuracy of test instruments must also be considered. Most voltmeters have accuracies to within a few percent (typically 5 to 10 percent). Precision laboratory meters (generally not used in troubleshooting) have much greater accuracy.

For proper operation, critical circuits may require that voltages be within a very close tolerance (at least within 10 percent and probably closer to 3 percent). However, many circuits will operate satisfactorily if the voltages are within 20 to 30 percent.

Generally, the most important factors to consider in voltage-measurement accuracy are the *symptoms* and the *output signal*. If no output signal is produced, you should expect a fairly large variation in voltages in the trouble area. Trouble that results in a circuit performance that is just out of tolerance may cause only a slight change in circuit voltages.

1-7.6 Resistance Measurements

Unlike voltage measurements, which are made with the equipment turned on, resistance measurements must be made with *no* power applied. However, in some cases various operating controls must be in certain positions to produce resistance readings similar to those found in the service literature charts. This is particularly true of controls that have variable resistances. Always observe any notes or precautions given in the charts. In any circuit, another safety precaution necessary to protect the ohmmeter is to be sure that all filter capacitors are *discharged.*

After these items are checked, the resistance measurements can be made from the terminals of the active device to the chassis (or ground) or between any two points that are connected by wiring or parts.

In military-type service literature, resistance information is presented in a format similar to that used for voltage information, as shown in Fig. 1-41. In some service literature, resistances are shown on the schematic diagram, along with voltage and wave-form information.

Do not be surprised if you find service literature with no resistance information whatsoever. The reason for this omission is fairly logical. If there is a condition in any circuit branch that will produce an abnormal resistance (e.g., a resistor is open, shorted, or has drastically changed value), the voltage at that circuit branch will be abnormal. If such an abnormal voltage reading is found, it is then necessary to check out each resistance in the circuit branch individually.

Because of the *shunting effect* of other components connected in *parallel,* the resistance of an individual component or circuit may be difficult to check. In such cases, it is necessary to disconnect one terminal of the component being tested from the rest of the circuit. This will leave the component open at one end, and the value of the resistance measured will be for that component only.

When making a resistance check, remember that a zero reading indicates a short circuit, and an infinity reading indicates an open circuit. Also remember the effect of the transistor junctions (acting as a forward-biased diode when biased on). The problems of resistance measurements are discussed thoroughly in Chapters 2 and 3.

1-7.7 Duplicating Wave-form, Voltage and Resistance Measurements

If you are responsible for the maintenance or service of one *piece* or one type of electronic equipment, it is strongly recommended that you use your own equipment to duplicate all the wave-form, voltage, and resistance measurements found in the service literature. This should be done with the equipment operating properly. Then, when you make measurements during troubleshooting, you can spot even slight variations in voltage. Always make the initial measurements with the test equipment that you will use during *actual* troubleshooting procedures. If you use more than one set of test equipment, make these initial measurements with all available test equipment, and *record the variations*.

1-7.8 Using Schematic Diagrams

Regardless of the type of trouble symptom, the actual fault can be traced eventually to one or more of the circuit components (vacuum tubes, transistors, ICs, diodes, resistors, capacitors, coils, transformers, and so forth). The checks of wave forms, voltage, and resistance will then indicate which branch within a circuit is at fault. Finally, you must locate the particular component that is causing the trouble in the branch.

In order to do this, you must be able to read a schematic diagram. These diagrams show what is inside the blocks on a servicing block diagram and provide the final picture of the electronic equipment. Often, you must service equipment with the aid of nothing more than a schematic diagram. If you are fortunate, the diagram will show some voltages and wave forms.

Figure 1-2 shows the schematic diagram of a solid-state radio receiver. This receiver unit (a portable broadcast radio) differs considerably from the one illustrated in the servicing block diagram shown in Fig. 1-19. For example, in the receiver shown in Fig. 1-2, there is no RF amplifier; the frequency-converter (mixer and RF oscillator) function is accomplished by one transistor, Q_1; there are only two IF amplifiers (one not shown); and there is no phase splitter in the audio section.

Examples of using schematic diagrams. The following examples demonstrate the use of the schematic diagram to locate a fault within a circuit. Although these examples involve a simple portable transistor radio, the same basic troubleshooting principles apply to more complex equipment.

In order to follow these examples properly, you must be able to use basic test equipment. If you are unsure of how such test equipment is used in troubleshooting, read Chapter 2 before you study these examples. (You may also read Chapter 2 in conjunction with these examples.)

Example 1. Assume that the receiver shown in Fig. 1-2 is being serviced and that the trouble is isolated to the frequency converter. This is done by injecting a signal of appropriate frequency (IF) modulated by an audio tone at the IF input (the primary winding of transformer T_3). A good response is heard on the loudspeaker. However, no response is obtained when a signal of proper frequency (RF) is injected at the primary winding of T_1 with the receiver tuned to that frequency.

The next step is to measure the voltages at the collector, emitter, and base of Q_1 (in that order), as shown in Fig. 1-42. If any of the Q_1 elements show an abnormal voltage, the resistance of the element should be checked first. Note

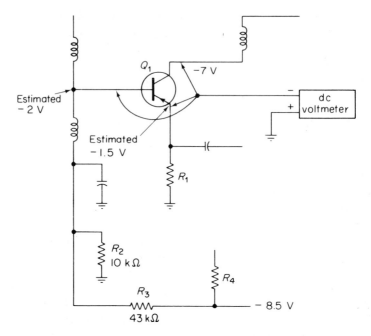

FIGURE 1-42   Measuring voltages at elements of Q_1

that the collector voltage is specified as -7 V. Neither the base and emitter voltages nor any of the resistance values are given. This lack of information is typical of electronic equipment used for home entertainment. Thus, you must be able to interpret schematic diagrams to estimate *approximate* voltages.

For example, the voltage at the junction of R_3 and R_4 is given as -8.5 V. This is logical because the source (a battery) is 9 V. The value of R_2 is approximately 25 percent of the value of R_3. Thus, the drop across R_2 is about 25 percent of -8.5 V, or *approximately* -2 V, and the base of Q_1 should be about -2 V. If Q_1 is silicon, the emitter will be about 0.5 V different from the base, or about -1.5 V (emitter more positive or less negative, in this case). If Q_1 is germanium, the base-emitter differential will be about 0.2 V, and the emitter should be approximately -1.8 V.

Keep in mind that this method of interpreting the schematic will give you *approximate* voltages only. In practical troubleshooting, the voltage *differentials* between circuit elements and transistor electrodes are the most important factor. Troubleshooting solid-state equipment based on voltage differentials is discussed fully in Chapter 3.

Now, assume that there is no voltage at the collector but that the base and the emitter show what appear to be normal or logical voltages. The next step is to remove power, discharge C_{14} and C_{15} (if necessary), and measure the

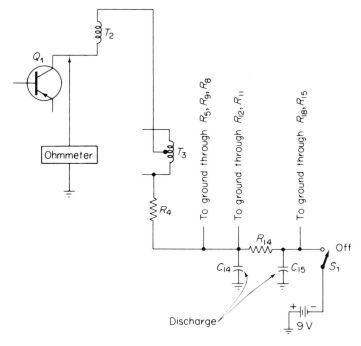

FIGURE 1-43 Measuring resistance at collector of Q_1

collector resistance (to ground), as shown in Fig. 1-43. Because no resistance values are given, you must use the schematic to estimate the approximate values.

Of course, if you find a zero resistance at the collector, this indicates a short. For example, capacitor C_5 could be shorted. On the other hand, an infinite resistance indicates an open circuit. For example, the coil windings of T_2 and T_3 could be open, or R_4 could be burned out and open. It is usually easy to locate the fault when you fiind such extreme resistance readings.

However, a resistance reading that falls between these two extremes does not provide a really sound basis for locating trouble. To make the problem worse, the effect of solid-state devices in the circuit can further confuse the situation.

For example, assume that the fault is an open T_3 winding, as shown in Fig. 1-44. This will result in a no-voltage reading at the collector of Q_1. However, it is still possible to measure a resistance to ground from the Q_1 collector if the following conditions are met: Assume that the ohmmeter leads are connected so that the *positive* terminal of the ohmmeter battery is connected to the Q_1 collector. This will forward-bias the automatic gain control (AGC) diode CR_1 and cause the ohmmeter to measure the resistance across R_7 (the collector supply of Q_2). Also, if the collector of Q_1 is made

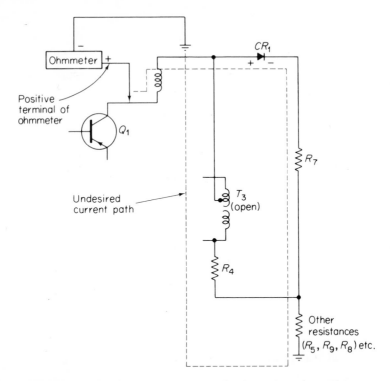

FIGURE 1-44   Undesired current path when measuring resistance at collector of Q_1

positive in relation to the base, the Q_1 base-collector junction will be forward-biased, resulting in possible current flow.

The problem illustrated in Fig. 1-44 can be eliminated by reversing the ohmmeter leads and measuring the resistance both ways. If there is a difference in the resistance values with the leads reversed, check the schematic for possible forward-bias conditions in diodes and transistor junctions in the associated circuit.

Example 2. Assume that the receiver shown in Fig. 1-2 is being serviced and that the trouble is isolated to the detector. An audio signal injected at the base of Q_3 produces a good response on the loudspeaker, as shown in Fig. 1-45. However, no response is obtained when an IF signal modulated by an audio frequency is injected at transformer T_4. Or using signal tracing, an oscilloscope connected across the secondary winding of T_4 shows that an AF modulated signal is available to the detector but that no AF signal appears at the base of Q_3.

The detector CR_2 has two outputs: an AF signal and an AVC voltage. Under no-signal conditions (i.e., no broadcast signal present), electrons flow

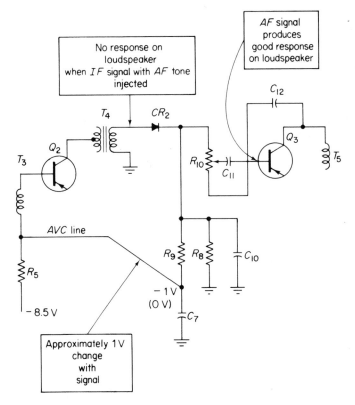

FIGURE 1-45   Example of troubleshooting detector-AVC circuits

through R_9 and R_8 to ground, producing approximately -1 V bias at the junction of R_5 and R_9. With a normal signal being received, electrons flow through R_8 and CR_2. This flow (opposite to that produced under no-signal conditions) opposes the no-signal bias and produces approximately 0 V at the junction of R_5 and R_9. This tends to reduce the gain of Q_2 (a *pnp* transistor) and thus tends to offset the signal passing through the IF stages.

The amount of AVC voltage is usually not critical, but the fact that the AVC voltage *changes* with changes in the IF signal is important. The function of the AVC circuit can be checked by measuring the voltage on the AVC line with and without a signal applied to T_4. If there is a change of approximately 1 V with signal, the AVC function and detector CR_2 can be considered normal.

With CR_2 established as normal but with no signal at the base of Q_3, volume control R_{10} or coupling capacitor C_{11} are logical suspects; they are probably open or have broken leads. Capacitor C_{12} can also be a possible suspect, but it is not as likely as C_{11} or R_{10}. If C_{12} is shorted or leaking, this will show up in an abnormal voltage (a large negative voltage on the AVC line). If C_{12} is open, the frequency-response characteristics of the audio circuit will be poor, but there will still be a signal at the base of Q_3.

The condition of C_{11} can be checked in two ways (without removal from the circuit). First, as shown in Fig. 1-46, a signal (audio frequency) can be injected on both sides of C_{11}. If the signal is heard on the loudspeaker when injected at the base of Q_3 but is not heard when injected at the junction of R_{10} and C_{11}, capacitor C_{11} is probably defective. Second, as shown in Fig. 1-47, a signal can be injected ahead of the detector and traced on both sides of C_{11} (with an oscilloscope or signal tracer). The signal should appear substantially the same on both sides of C_{11}. If not, C_{11} is probably defective.

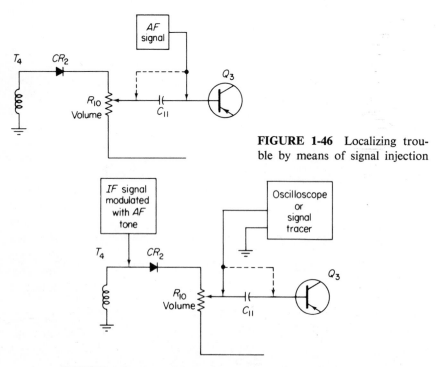

FIGURE 1-46 Localizing trouble by means of signal injection

FIGURE 1-47 Localizing trouble by means of signal tracing

The condition of R_{10} can also be checked by signal tracing or signal injection. However, as a final test, the power should be removed and the resistance of R_{10} measured. To measure potentiometer resistance, connect the ohmmeter from the wiper arm to one end of the winding, as shown in Fig. 1-48. Then, vary the control from one extreme to the other. Repeat the test with the ohmmeter connected from the wiper arm to the opposite end of the winding. The resistance indication should vary smoothly, with no jumps or flicking of the ohmmeter needle (or jitters of the display if a digital ohmmeter

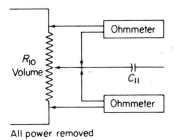

FIGURE 1-48 Measuring resistance of volume control R_{10} All power removed

is used). Such jumps can mean bad spots (open or poor contact) on the potentiometer winding.

1-7.9 Internal Adjustments during Trouble Localization

Keep in mind that adjustment of controls (both internal-adjustment controls and operational controls) can affect circuit conditions. This may lead to false conclusions during troubleshooting. For example, the amplitude of the signal at the base of Q_3 (Fig. 1-2) is set directly by the volume control R_{10}, which is an operational control. The amplitude of both the signal and the AVC voltage from the detector can be affected by adjustment (or alignment) of the IF transformers (as well as the oscillator and the RF transformers).

If the signal at Q_3 is very low, it could be that the volume control is set to the minimum position. Of course, because the volume control is an operational control, a run-through of the operating sequence at the beginning of troubleshooting will pinpoint such an obvious condition. However, a low output from the detector can be caused by poor alignment of the IF transformers. Because the IF transformers require internal adjustments, poor alignment will not be detected through the use of operating procedures. This condition, or a similar one, could lead you to one of two unwise courses of action.

First, you might launch into a complete alignment procedure (or whatever internal adjustments are available) once you have isolated the trouble to a circuit and are trying to locate the specific defect. No internal control, no matter how inaccessible, is left untouched. You reason that it is easier to make adjustments than to replace parts. Such a procedure will eliminate improper adjustment as a possible fault, but it can also create more trouble than it repairs. Indiscriminate internal adjustment is the technician's equivalent of operator trouble.

Second, you might replace part after part when a simple screwdriver adjustment would repair the problem. This attitude is usually caused by the inability to perform the adjustment procedures or by a lack of knowledge concerning the control's function in the circuit. Either way, a study of the service literature should resolve the situation.

But there *is* a middle ground. Do not make any internal adjustments during the troubleshooting procedure until trouble has been isolated to a circuit, and then only when the trouble symptom or test results indicate possible maladjustment.

For example, assume that an oscillator is provided with an internal-adjustment control that sets the frequency of oscillation. If wave-form measurement at the circuit shows that the oscillator is off frequency, it is logical to adjust the frequency control. However, if wave-form measurement shows only a very low output (but on frequency), adjustment of the frequency control during troubleshooting could cause further problems.

An exception to this rule occurs when the service literature recommends alignment or adjustment as part of the troubleshooting procedure. Generally, alignment or adjustment is checked after testing and repairs have been completed. This assures that the repair procedure (replacement of parts) has not upset circuit adjustment.

1-7.10 Trouble Resulting from More than One Fault

A review of all the symptoms and test information obtained thus far will help you to verify the component located as the sole trouble or to isolate other faulty components. This is true whether the malfunction of these components is caused by the isolated component or by some entirely unrelated problem.

If the isolated malfunctioning component can produce all the normal and abnormal symptoms and indications that you have accumulated, you can logically assume that it is the sole cause of the trouble. If it cannot, you must use your knowledge of electronics and of the equipment to determine what other component or components could have become defective and produced all the symptoms.

When one component fails, it often causes abnormal voltages or currents that could damage other components. Trouble is often isolated to a faulty component that is a result of the original trouble rather than its source.

For example, assume that the troubleshooting procedure thus far has isolated a transistor at the cause of trouble and that the transistor is burned out. What would cause this? Excessive current can destroy the transistor by causing internal shorts or by altering the characteristics of the semiconductor material, which is very sensitive to temperature. Thus, the problem becomes a matter of finding how excessive current can be produced.

Excessive current in a transistor can be caused by an extremely large input signal, which will overdrive the transistor. Such an occurrence will indicate a fault somewhere in the circuitry ahead of the input connection. Power surges (intermittent excessive outputs) from the power supply can also cause the transistor to burn out. In fact, power supply surges are a common cause of transistor burnout. All these conditions should be checked *before* a new transistor is placed in the circuit.

Some other typical malfunctions, along with their common causes, include:

1. Burned-out transistors caused by *thermal runaway.* An increase in transistor current heats the transistor, causing a further increase in current, resulting in more heat. This continues until the heat-dissipation capabilities of the transistor are exceeded. Bias-stabilization circuits are generally included in most well-designed transistor equipment.

2. Power supply overload caused by a short circuit in some portion of the voltage-distribution network.

3. Burned-out transistor in shunt-feed system caused by shorted blocking capacitor.

4. Blown fuses caused by power supply surges or shorts in filtering (power) networks.

It is obviously impractical to list all the common malfunctions and their related causes that you may find in troubleshooting electronic equipment. Generally, when a component fails, the cause is an operating condition that exceeded the maximum ratings of the component. However, it is quite possible for a component simply to "go bad."

The operating condition that causes a failure can be temporary and accidental. Or it can be a basic design problem, as a history of repeated failures would indicate. No matter what the cause, your job is to find the trouble, verify its source or cause, and then repair it.

1-7.11 Repairing Troubles

In a strict sense, repairing the trouble is not part of the troubleshooting procedure. However, repair is an important part of the total effort involved in getting equipment back into operation. Repairs must be made before the equipment can be checked out and declared ready for operation.

Never replace a component if it fails a second time unless you make sure that the cause of trouble is eliminated. Actually, the cause of trouble should be pinpointed *before* you replace a component the first time. However, this is not always practical. For example, if a resistor burns out because of an intermittent short and you have cleared the short, the next step is to replace the resistor. However, the short could recur and burn out the replacement resistor. If this happens, you must recheck every element and lead in the circuit.

When replacing a defective component, an *exact replacement* should be used if it is available. If an exact replacement is not available and the original component is beyond repair, an *equivalent or better* component should be used. *Never* install a replacement component that has characteristics or ratings inferior to those of the original. This problem is discussed further in Chapter 3.

An exception to this rule may be made when an equivalent or better component is not available and it is imperative that the equipment be placed in operation in the shortest possible time. Of course, once the emergency is over, an equivalent or better component must be installed as soon as it becomes available.

Another factor to consider when repairing the trouble is that the replacement component should be installed in the *same physical location* as the original, with the same lead lengths and so on, if at all possible. This precaution is optional in most low-frequency or dc circuits, but it must be followed for high-frequency applications.

In high-frequency (RF, IF, video, and so forth) circuits, changing the location of components or the length of leads may detune the circuit or otherwise put it out of alignment.

1-7.12 Operational Checkout

Even after the trouble has been found and the faulty component located and replaced, the troubleshooting effort is not necessarily completed. An operational check must be performed to verify that the equipment is free of *all* faults and is performing properly again. Never assume that simply because a defective component has been located and replaced, the equipment will automatically operate normally again. As a matter of fact, in practical troubleshooting of any electronic equipment, never *assume* anything; *prove* it!

Operate the equipment through its *complete* operating sequence. In this way, you will make sure that one fault has not caused another. Follow the procedure found in the service literature (when available). In the case of specialized equipment, have the operator go through the entire sequence, but verify operation yourself.

When the operational check is completed and the equipment is again certified to be operating normally, make a brief record of the symptoms, the faulty component, and the remedy. This is particularly helpful when you must service the same equipment on a regular basis or when you must troubleshoot similar equipment. Even a simple record of troubleshooting will give you a valuable history of the equipment for future reference.

If the equipment does not perform properly during the operational checkout, you must continue troubleshooting. If the symptoms are the same as, or similar to, the original trouble symptoms, retrace your steps *one at a time*. If the symptoms are entirely different, you may have to repeat the entire troubleshooting procedure from the beginning. However, this is usually not necessary.

For example, assume that the equipment does not check out because a replacement IF amplifier transistor has detuned the circuit. In this case, you should repair the trouble by IF alignment rather than by returning to the first troubleshooting step and repeating the entire procedure. Keep in mind that you have arrived at the defective circuit or component by a systematic procedure. Therefore, retracing your steps—one at a time—is the logical course of action.

2

Test Equipment Used
for Basic Troubleshooting

In this chapter, we shall discuss the basic test equipment used in troubleshooting. We describe, in basic terms, what equipment is available, how it works, how to operate it for its basic functions, and how to use it in troubleshooting.

A thorough study of this chapter will make you familiar with the basic principles and operating procedures for equipment used in general circuit troubleshooting. It is assumed that you will take the time to become equally familiar with the principles and operation controls for any *particular* test equipment you use. Such information is contained in the service literature for the particular equipment.

It is absolutely essential that you become thoroughly familiar with your particular test instruments. No amount of textbook instruction will make you an expert in operating test equipment; it takes actual practice.

It is strongly recommended that you establish a *routine* operating procedure or sequence of operation for each item of troubleshooting test equipment. This will save time and familiarize you with the capabilities and limitations of your particular equipment, thus eliminating false conclusions based on unknown operating conditions.

2-1. SAFETY PRECAUTIONS IN TROUBLESHOOTING

In addition to establishing a routine operating procedure, you must observe certain precautions during operation of any electronic test equipment. Many of these precautions are the same for all types of test equipment; others are unique to special test instruments, such as meters, oscilloscopes, and signal generators. Some of the precautions are designed to prevent damage to the test equipment or to the circuit where the troubleshooting operation is being performed. Other precautions are to prevent injury to the troubleshooter. Where applicable, special safety precautions are noted throughout this book.

The following general safety precautions should be studied thoroughly and then compared with any specific precautions called for in the service literature of a particular piece of equipment and in the related chapters of this book.

1. Many troubleshooting instruments are housed in metal cases. These cases are connected to the ground of the internal circuit. For proper operation, the ground terminal of the test instrument should always be connected to the ground of the equipment being serviced. Make certain that the chassis of the equipment being serviced is not connected to either side of the ac line or to any potential not at ground. If there is any doubt, connect the equipment being serviced to the power line through an *isolation transformer*.

2. Remember that troubleshooting equipment that operates at hazardous voltages is always dangerous. Therefore, you should familiarize yourself thoroughly with the equipment being serviced before troubleshooting it, bearing in mind that high voltage may appear at unexpected points in defective equipment.

3. It is good practice to remove power *before* connecting test leads to high-voltage points. It is preferable to make all troubleshooting connections with the power removed. If this is impractical, be especially careful to avoid accidental contact with equipment and objects that are grounded. If you work with one hand away from the equipment and stand on a properly insulated floor, you will lessen the danger of electric shock.

4. Capacitors may store a charge large enough to be hazardous. Discharge filter capacitors before attaching test leads.

5. Remember that leads with broken insulation pose the additional hazard of high voltages appearing at *exposed* points along the leads. Check test leads for frayed or broken insulation before working with them.

6. To lessen the danger of accidental shock, disconnect test leads *immediately after* the test is completed.

7. Remember that the risk of severe shock is only one possible hazard. Even a minor shock can place you in danger of more serious risks, such as a bad fall or contact with a source of higher voltage.

8. Guard continuously against injury and do not work on hazardous circuits unless another person is available to assist you in case of accident.

9. Even if you have had considerable experience with test equipment used in troubleshooting, always study the service literature of any instrument with which you are not thoroughly familiar.

10. Use only shielded leads and probes. Never allow your fingers to slip down to the metal probe tip when the probe is in contact with a hot circuit.

11. Avoid vibration and rough treatment. Most electronic test equipment is delicate.

12. Study the circuit being serviced before making any test connections. Try to match the capabilities of the test instrument to the circuit being serviced.

2-2. ANALOG METERS

Most meters used in troubleshooting are *analog meters*. That is, such meters use rectifiers, amplifiers, and other circuits to generate a current *proportional* to the quantity being measured. This current, in turn, drives a meter movement or a digital readout. (The digital meters described in later sections are of the analog type, but they differ from the mechanical-movement type in the manner of readout.)

Although both commercial and laboratory troubleshooting meters use the same basic analog principles, laboratory-type meters include many circuit refinements to improve their accuracy and stability.

This section describes the basic principles of analog meters and shows how these principles are adapted to troubleshooting.

2-2.1 Meter Basics

The main purpose of any meter used in troubleshooting is to check circuits and components (i.e., to find what voltage is available, how much current is flowing, and so forth). The simplest and most common instrument that will measure the three basic electrical values (voltage, current, and resistance) is the *volt-ohm-milliammeter* (VOM). The VOM is the basic troubleshooting instrument, and there are dozens available in all price ranges. The more expensive VOMs have greater accuracy, more scales or functions, and greater scale range.

The first improvement on the VOM was the *vacuum-tube voltmeter* (VTVM). Today, the VTVM has generally been replaced by *electronic voltmeters* (EVMs), such as the *transistorized voltmeter* (TVM) and the *field-effect* meter. The sensitivity of the EVM is much greater than that of the VOM because there is a transistor amplifier between the meter readout and the input. The EVM has another advantage over the VOM in that it presents a *high impedance* to the circuit or component being measured. Thus, EVMs draw little current from the circuit and have little effect on circuit operation.

2-2.2 Basic VOM

Except in the case of digital-readout meters (discussed in Sec. 2-3), all meter circuits are designed around the basic meter movement. Virtually all nondigital meters use some form of the D'Arsonval meter movement, as shown in Fig. 2-1. This movement is also known as the *moving-coil galvanometer*. By itself, the basic movement forms an *ammeter* (*am*pere *meter*). However, when the movement is used in a VOM, other circuit components are added to extend the range and to perform *voltmeter* and *ohmmeter* functions.

Basic ammeter. A true ammeter measures current in amperes. In electronic troubleshooting, current is more often measured in milliamperes (mA) or microamperes (μA). Most movements used in VOMs will produce full-scale deflection when 50 μA is passed through them.

A *shunt* must be connected across the meter movement if currents greater than the full-scale range of the basic movement are to be measured.

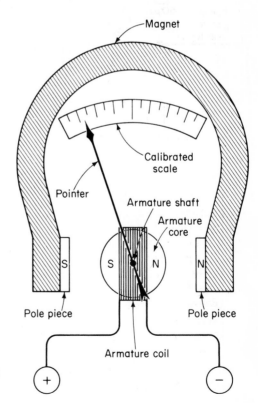

FIGURE 2-1   Basic D'Arsonval meter movement

The shunt can be a precision resistor, a bar of metal, or a simple piece of wire. VOM shunts are usually precision resistors that may be selected by means of a switch. Shunt resistance is only a fraction of the movement resistance. Current divides itself between the meter and the shunt, with most of the current flowing through the shunt. Shunts must be precisely calibrated to match the meter movement.

Figures 2-2 and 2-3 show two typical ammeter range-selection circuits for VOMs. In Fig. 2-2, individual shunts are selected by the range-scale selector. In Fig. 2-3, the shunts are cut in or out of the circuit by the selector. If the selector is in position 1, all three shunts are across the meter movement, giving the least shunting effect (most current through the movement). In position 2, resistor R_1 is shorted out of the circuit, with resistors R_2 and R_3 shunted across the movement, increasing the meter's current range. In position 3, only R_3 is shunted across the movement, and the meter reads maximum current.

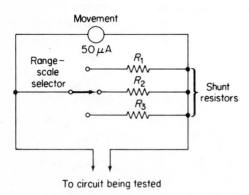

FIGURE 2-2 Typical ammeter range-scale selector circuit (individual-shunt method)

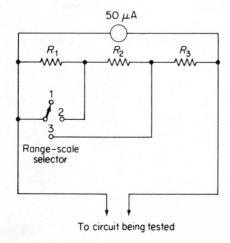

FIGURE 2-3 Typical ammeter range-scale selector circuit (series-shunt method)

Basic voltmeter. When the basic meter movement is connected in series with resistors, a voltmeter is formed. The series resistance is known as a *multiplier* because the resistance multiplies the range of the basic meter movement.

The basic voltmeter circuit is shown in Fig. 2-4. The voltage divides itself across the meter movement and the series resistance. If a 0.5 V full-scale meter movement is used and a full scale of 10 V is to be measured, the series resistor must drop 9.5 V, if a 100 V full-scale is used, the series resistance must drop 99.5 V; and so on.

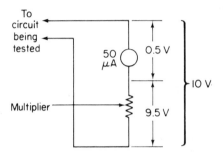

FIGURE 2-4 Basic voltmeter circuit

Figures 2-5 and 2-6 show the two typical voltmeter range-selection circuits for VOMs. In Fig. 2-5, individual multipliers are selected by the range-scale selector. In Fig. 2-6, the multipliers are cut in or out of the circuit by the selector. If the selector is in position 1, only resistor R_1 is in the circuit, giving the least voltage drop (meter will read the lowest voltage). In position 2, both R_1 and R_2 are in the circuit, giving the meter a higher voltage range. In position 3, all three resistors drop the voltage, permitting the meter to read the maximum voltage.

The measurement *ohms per volt* (Ω/V) is used in connection with commercial VOMs. It is a measure of the sensitivity of a VOM and represents the number of ohms required to extend the range by 1 V. For example, if the meter movement requires 1 mA for full-scale deflection, then 1,000 Ω

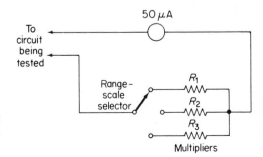

FIGURE 2-5 Typical voltmeter range-selector circuit (individual-multiplier method)

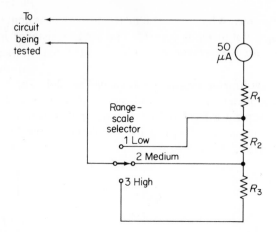

FIGURE 2-6 Typical voltmeter range-selector circuit (series-multiplier method)

(including the movement's internal resistance) are needed for each volt that could be measured if the movement is used as a voltmeter. If the movement requires only 100 μA for full-scale deflection, then 10,000 Ω/V are needed. Thus, the more sensitive the meter movement (those that require the least current for full-scale deflection), the higher the ohms-per-volt rating.

Voltmeters with a high ohms-per-volt rating put less load on the circuit being measured and have a less disturbing effect on the circuit. For example, assume that a 1 V drop across a 1,000 Ω circuit is to be measured with both a 1,000 Ω/V meter and a 20,000 Ω/V meter.

A 1 V drop across a 1,000 Ω circuit will produce a 1 mA current flow. With the 1,000 Ω/V meter across the circuit, the 1 mA current will divide itself between the meter and the circuit. This will cut the circuit's normal current in half. With a 20,000 Ω/V meter across the same circuit, one-twentieth of the current will pass through the meter, and nineteen-twentieths will remain in the circuit.

Basic ohmmeter. An ohmmeter (or resistance-measuring device) is formed when a basic meter movement is connected in series with a resistance and a power source (such as a battery). The basic ohmmeter arrangement is shown in Fig. 2-7. Here, a 3 V battery is connected to a meter movement with a full-scale reading of 5 mA. The value of the current-limiting resistor R (600 Ω, less the meter resistance) is such that exactly 5 mA will flow in the circuit when the test leads are clipped together.

When there is no connection across the test leads, the current will be zero. The meter's pointer will rest at the INFINITY mark on the scale. (The symbol for infinity is ∞.) When the two leads are shorted, the meter will

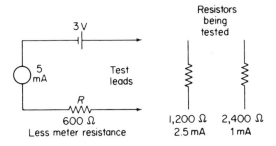

FIGURE 2-7   Basic ohmmeter circuit

move to its full 5 mA reading, which will indicate on the scale that there is zero resistance at the test leads.

If a 600 Ω resistance is connected across the leads, the total resistance is 1,200 Ω, and the meter will drop to one-half in full-scale reading, or 2.5 mA.

If the battery voltage and limiting resistor R remain constant, the pointer will always move to 2.5 mA whenever a 600 Ω resistance is connected across the test leads. The 2.5 mA point on the meter scale is marked 600 Ω.

With a 2,400 Ω resistance across the leads, the total resistance is 3,000 Ω, and the point will drop to a 1 mA reading because $I = E/R$, or 3/3,000 = 0.001 A (1 mA).

Again, if the battery voltage and the limiting resistor remain constant, the meter will always read 1 mA when a resistance of 2,400 Ω is placed across the test leads. Thus, the 1 mA point on the meter scale is marked 2,400 Ω.

The ohmmeter arrangement shown in Fig. 2-7 is thus capable of measuring 600 Ω and 2,400 Ω. Any number of resistance values can be plotted on the scale, provided resistances of known value are placed across the leads.

The scale of a commercial VOM has its own markings or calibrations. As discussed later in this section, the ohmmeter is printed on the meter face along with the voltage and current scales. However, the ohmmeter scale is quite different from the other scales in two respects: The *zero point* is at the right-hand side (usually), and the *maximum resistance* (usually marked INFINITY) is at the left-hand side. Also, the scale is not linear (i.e., lower-resistance divisions are wider, and higher-resistance divisions are narrower).

The ohmmeter circuit of a typical VOM is shown in Fig. 2-8. The ohmmeter has two range scales that can be selected by means of a switch. In the HIGH position, a series multiplier (similar to that of a voltmeter) is connected to the circuit and drops the voltage by a corresponding amount. This reduces current flow through the entire circuit, usually by a ratio of 10:1, 100:1, or 1,000:1, so that the ohmmeter scale represents 10, 100, or 1,000 times the indicated amount. In the case of the ohmmeter circuit (Fig. 2-7)

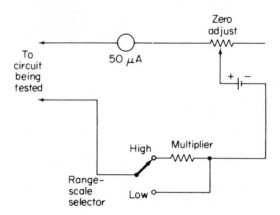

FIGURE 2-8   Ohmmeter circuit of typical VOM

just described, the 600 Ω point (half scale) would represent 6,000, 60,000, or 600,000 Ω.

No matter which scale is used, the meter and battery are in series with a variable resistor that allows the circuit to be zeroed. As a battery ages, its output drops. Also, it is possible that the resistance values (or the meter movement itself) could change because of extended age or extreme temperature. Any of these conditions will make the ohmmeter scale inaccurate. The variable resistor (usually marked ZERO ADJUST or ZERO) is included in a commercial VOM circuit. In use, the leads are shorted together, and the variable resistor is adjusted until the meter is at ZERO. When the leads are opened, the meter then moves back to INFINITY or OPEN, and the meter is ready to read resistance accurately.

Basic galvanometer. The basic meter movement described thus far can be used as a galvanometer. However, the term *galvanometer* has come to mean a meter where the ZERO of the scale is at the *center*, with negative-current reading to the left and positive-current reading to the right, as shown in Fig. 2-9. Generally, a galvanometer is used to read *proportional* positive or negative changes in circuits rather than the actual unit value of current. The main use for such a meter is in a *bridge circuit* (described in Sec. 2-4).

2-2.3 Basic AC Meters

Most ac meters are similar to dc meters in that they are analog current-measuring devices. However, the basic meter movement cannot be connected directly to alternating current because it reverses direction during each cycle. Instead, the meter movement is connected to the ac voltage through a *rectifier*. Both half-wave and full-wave rectifiers are used. However, the full-wave bridge rectifier shown in Fig. 2-10 is most efficient because a direct

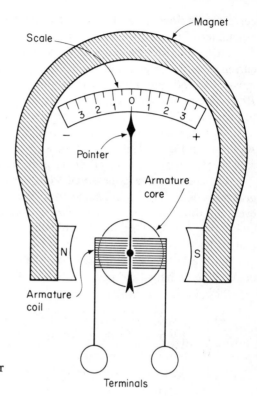

FIGURE 2-9 Basic zero-center galvanometer movement

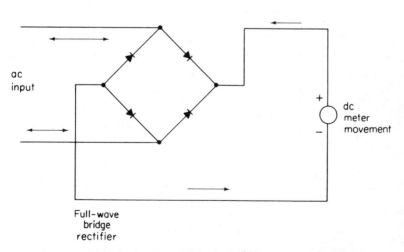

FIGURE 2-10 Basic ac meter circuit with full-wave bridge rectifier

current will flow through the meter movement on both half-cycles. The remainder of the ac meter circuit can be identical to that of a dc meter.

Such an arrangement will work well with low-frequency alternating currents, but it presents a problem as frequency increases. This is because the movement and multiplier resistances may load the circuit being tested. An *RF probe* can be connected between the meter and the circuit being measured to overcome this condition. (Meter probes are discussed in Sec. 2-7.)

Ac meter scales also present a problem that does not occur on dc scales. As shown in Fig. 2-11, there are four ways to measure an ac voltage: average, root-mean-square (rms) or effective, peak, or peak to peak.

The *peak* voltage is measured from the crest of one half-cycle; whereas *peak-to-peak* voltage is measured from the crests of both half-cycles. However, the direct current to the meter movement will be less than the peak alternating current because the voltage and current drop to zero on each half-cycle.

With a full-wave bridge rectifier, the current or voltage is 0.637 of the peak value (a half-wave rectifier delivers 0.318 of the peak value). This is

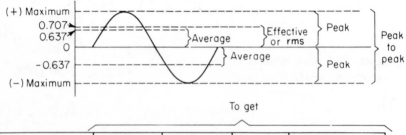

To get

Given	Average	Effective (rms)	Peak	Peak to peak
Average	——	1.11	1.57	1.271
Effective (rms)	0.900	——	1.411	2.831
Peak	0.637	0.707	——	2.00
Peak to peak	0.3181	0.3541	0.500	——

FIGURE 2-11   Relationship of average, effective or rms, peak, and peak-to-peak values for ac sine waves

known as *average* value, and some meters are so calibrated. Most meters have rms scales. On an rms meter, the scale indicates 0.707 of the peak value (assuming that the usual full-wave rectifier is used). This value is the *effective* value of an alternating current.

In service literature used for troubleshooting, most ac voltage information in given in terms of rms value. When peak voltages are involved, as they are in wave-form measurements, they are measured on an oscilloscope (Sec. 2-6) rather than a meter.

Current probes and clip-on meters. Most troubleshooting procedures can be accomplished without measuring currents. However, when ac currents are to be measured, a clip-on meter or current probe can be used, as shown in Fig. 2-12. Alternating currents can be picked up by a coil of wire around the conductor, stepped up through a transformer, and measured by a voltmeter.

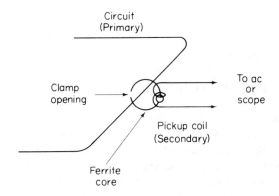

FIGURE 2-12   Basic clip-on meter or current probe circuit

A *clip-on meter*, complete with built-in coil, transformer, and meter movement, is particularly useful if conductors are carrying heavy currents and if it is not convenient to open the circuit to insert an ammeter. Clip-on meters are most often used for heavy industrial work.

A *current probe* is similar to a clip-on meter except that it can be used in conjunction with an amplifier to measure small current in troubleshooting. In a commercial current probe, the amplifier is adjusted so that 1 mA in the wire being measured produces 1 millivolt (mV) at the amplifier output. In this way, current can be read directly on the voltmeter.

2-2.4 Electronic Analog Meters

Electronic analog meters operate by producing a current proportional to the quantity being measured, as do basic VOMs. However, electronic meters include many circuit refinements to improve their accuracy and stability. Also, there are special-purpose circuits that are unique to electronic meters.

The basic electronic meter circuit is shown in Fig. 2-13. The amplifier can be either vacuum tube or transistor and is usually direct coupled. Field-effect transistors (FETs) are often used for electronic meter amplifiers because FETs have a high input impedance.

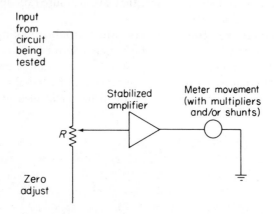

FIGURE 2-13   Basic electronic voltmeter circuit

Electronic voltmeter. When the basic circuit is used as a voltmeter, resistance R has a large value (usually several megohms) and is connected in parallel across the voltage circuit being measured, as shown in Fig. 2-14. Because of the high resistance, very little current is drawn from the circuit, and operation of the circuit is not disturbed. The voltage across resistance R raises the voltage at the amplifier input from the zero level. This causes the meter at the amplifier output to indicate a corresponding voltage.

Electronic ammeter. When the basic circuit is used as an ammeter, resistance R has a small value (a few ohms or less) and is connected in series

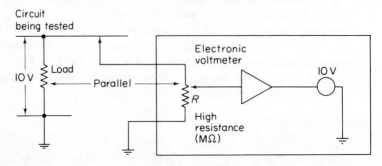

FIGURE 2-14   Basic electronic voltmeter measurement connections

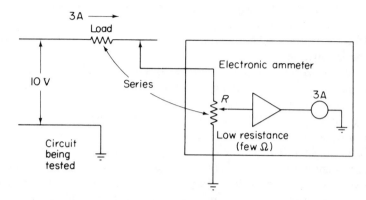

FIGURE 2-15   Basic electronic ammeter measurement connections

with the circuit being measured, as shown in Fig. 2-15. Because of the low resistance, there is little change in the total circuit current, and operation of the circuit is not disturbed. Current flow through resistance R causes a voltage to be developed across R, which raises the level at the amplifier input from zero and causes the meter to indicate a corresponding (current) reading.

Differential amplifier. One of the most common circuits used in electronic meters is shown in Fig. 2-16. This is essentially a differential amplifier, in which the voltage to be measured is applied to one input and the other input is grounded. The zero-set resistance is adjusted so that the meter reads ZERO when no input voltage is applied. When the voltage to be measured is applied across the input resistance, the circuit is unbalanced, and the meter indicates the proportional imbalance as a corresponding voltage reading.

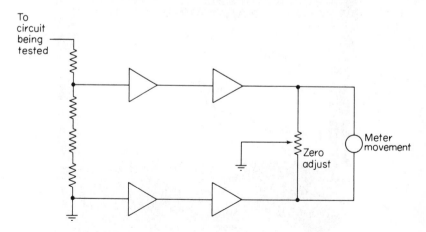

FIGURE 2-16   Basic electronic voltmeter with differential amplifier

One of the reasons for using the differential-amplifier circuit is to minimize drift resulting from power supply changes.

Meter amplifiers. The amplifier shown in Figs. 2-13 and 2-16 performs three basic functions.

First, the effective *sensitivity* of the meter movement is increased. An amplifier changes the measured quantity to a current of sufficient amplitude to deflect the meter movement. Thus, a few microvolts (μV) that would not show up on a typical VOM can be amplified to several volts to deflect any meter movement.

Second, the amplifier increases the *input impedance* of the meter so that the instrument draws little current from the circuit under test. This is critical when you are troubleshooting many sensitive circuits. A typical electronic meter will provide an 11 MΩ input impedance. If an FET is used in the meter's input circuit, the input is increased to about 100 MΩ.

Third, the amplifier limits the *maximum current* applied to the meter movement. Therefore, there is little danger when unexpected overloads occur that could burn out the meter movement.

2-2.5 Meter Scales and Ranges

Figures 2-17 and 2-18 show the operating controls and scales for a typical VOM and a typical electronic volt-ohmmeter, respectively. This

FIGURE 2-17   Simpson Instruments Model 262 VOM

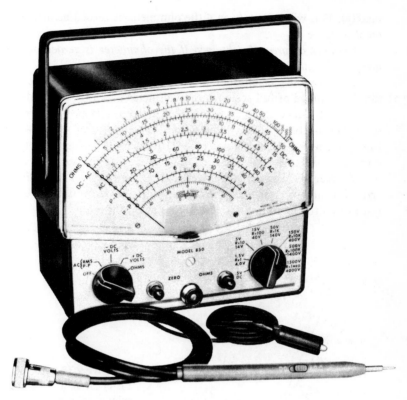

FIGURE 2-18   Triplett Model 850 electronic volt-ohmmeter

section describes each of the scales and provides information concerning their accuracy and use during troubleshooting.

Ohmmeter scales. Note that the zero indication is at the right on the VOM ohmmeter scale and at the left on the electronic ohmmeter scale. Although this is typical, it will not be found on every make and model of meter.

Also note that the high-resistance end (or INFINITY end) of the ohmmeter scale is cramped on both meters. Ohmmeter scales are always nonlinear. Therefore, ohmmeters provide their most accurate indications at mid-scale or near the low-resistance end.

In general, the ohmmeter scale is considered to be as accurate as the dc voltmeter scale. However, in the case of a battery-operated VOM, the condition of the battery will affect accuracy. As a battery ages and its voltage output drops, the resistance indications will be *lower* than the actual value. You should consider this point when making resistance measurements during troubleshooting. For example, if a battery voltage drops to 90 percent of its minimum value, a 100 Ω resistor will produce a 90 Ω indication (approximately). This will be true even if the ohmmeter is zeroed before the measurement is made.

Greatest accuracy is obtained if the ohmmeter is zeroed on each range just prior to making the measurement.

On some meters, the ohmmeter scale is rated in *degrees of arc* rather than percentage of full scale (as the voltmeter and ammeter scales are). For example, the dc voltage accuracy of a meter could be ±3 percent of full scale. On the 100 V scale, this indicates an accuracy of ±3 V. As shown in Fig. 2-19, a ±3 V (or a total of 6 V) indication corresponds to a certain number of degrees of arc. In turn, this arc defines the accuracy of the ohmmeter scale. Because the ohmmeter scale is nonlinear, it is possible that the error will not be constant over the entire scale. However, the error should not exceed the rated accuracy at any point on the scale.

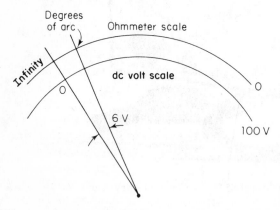

FIGURE 2-19   Accuracy of ohmmeter scale (in degrees of arc) related to accuracy of dc voltage scale (in percentage of full scale)

dc scales. The ZERO indication for the dc scales is almost always on the left. Usually, the dc scales are linear, with no cramping or bunching at either end. Note that there are three basic dc scales (with maximums of 8, 40, and 160) on the VOM (Fig. 2-17). Each of these scales serves many purposes, depending on the position of the range selector. (The range selected is indicated by a dot pointer at the scale center.) Therefore, you must make note of *both* the scale reading and the range-selector position before you can obtain the correct indication.

For example, the dc readings of 60, 15, and 3 are all aligned. These readings are located on the 160, 40, and 8 scales, respectively. If the range selector is set to 1.6 V dc, then the 60 reading applies, and the indicated value is 0.6 V dc. If the readings are the same but the range selector is set to 160 mA, then the 60 reading still applies, but the indicated value is 60 mA. Again, with the same reading but with the range selector set to 400 V dc, the 15 reading applies, and the indicated value is 150 V dc.

The use of meter scales is often confusing to the inexperienced trouble-shooter. Therefore, you should study the meter scales on your particular equipment thoroughly before attempting to use the meter in troubleshooting.

The accuracy of the dc scales is dependent upon the tolerance of the multiplier resistors, the accuracy of the movement, and the accuracy of the scales. In precision meters, the scales are matched to individual movements so that the combined movement and scale has a given accuracy. Usually, the multiplier (for voltage) and shunt resistors (for current) have an accuracy of ± 1 percent or better. When this is coupled to a movement with a ± 2 percent or better accuracy, the total accuracy is ± 3 percent.

The accuracy of both dc and ac scales is usually specified as a *percentage of full scale*, rather than the actual reading. This is another fact that the inexperienced troubleshooter often overlooks when making voltage readings.

For example, assume that a voltage is measured during troubleshooting on the 40 V dc scale (Fig. 2-17), that a reading of 10 is obtained, and that the rated accuracy is ± 3 percent of full scale. The full-scale value is 40 V; therefore, the absolute accuracy is ± 1.2 V (40 $\times$ 3 percent). The 10 V reading could indicate an actual value of from 8.8 to 11.2 V. Inexperienced troubleshooters often assume (incorrectly) that the 3 percent tolerance applies directly to the reading (or 0.3 V in this case). To make sure, always consult the meter's service literature to determine how accuracy is specified.

Differences in dc scales. Several differences are noted between the dc scales and ranges of the VOM (Fig. 2-17) and those of the electronic volt-ohmmeter (Fig. 2-18).

First, the electronic volt-ohmmeter is provided with both a *zero adjustment* and an ohms adjustment. Usually, this ZERO control must be set so that the movement's needle is aligned with zero each time the range selector is moved to another range position.

There are several reasons for this. In vacuum-tube meters, there is a contact potential at the tube elements (particularly the grid) that changes with each position of the range selector. In transistor meters, there is a certain amount of current leakage that is subject to change. Also, there is a possibility of drift due to temperature change during warm-up or over a long period of operation. All these conditions are compensated for by the ZERO control, which is set to provide meter zero each time the range selector is moved and periodically thereafter throughout the operation.

Second, note that the meter's function switch shows both a *minus* dc volts position and a *plus* dc volts position. In a simple VOM, positive and negative voltages are usually adjusted for by *reversing the leads*. Usually, the black lead is connected to negative ($-$), and the red lead is connected to positive ($+$) when a positive dc voltage is to be read. The leads are then reversed for a negative dc voltage. Thus, the voltage indication is also a *polarity* indication. (This should be verified, however, by consulting the meter's service literature. If the literature is not available, check for correct lead connections with a battery or some other dc source with a known polarity.)

On the electronic volt-ohmmeter, the alligator clip of the common lead is connected to ground, and the circuit voltage is measured with the probe tip. The test leads are *never* reversed. Thus, it is necessary to have a negative and a positive position for measurement of dc voltages. On some electronic volt-ohmmeters, the probe rather than the meter itself is provided with a switch for reversing polarity.

Third, note that the electronic volt-ohmmeter does not have any current-measuring function. This is common for most such meters in general use. However, certain electronic volt-ohmmeters will measure current.

Fourth, note that the electronic volt-ohmmeter has a special *zero-center* scale (bottom scale). This scale permits both positive and negative-voltage indications to be displayed on either side of a zero center without reversing the function switch or the test leads.

In use, the function switch is set to + dc volts and the ZERO knob is adjusted until the meter needle is aligned with the zero center (with no voltage applied). A positive voltage will deflect the needle to the right (+), and a negative voltage will move the needle to the left (−).

On most electronic volt-ohmmeters, the divisions of the zero-center scale are arbitrary and have no relation to actual voltage. However, the divisions do provide a relative measure of voltage on either side of zero. The zero-center scale permits the meter to be used as a null meter in a bridge circuit, as an alignment indication tool for the detector of an FM receiver, and so on.

ac scales. The zero indication for ac scales is almost always on the left. Usually, the ac scales are somewhat nonlinear, with some cramping or bunching on the low end. This is because the rectifier circuits required for ac measurement are nonlinear. Both half-wave and full-wave rectifier circuits are nonlinear. This nonlinearity is more pronounced when the multiplier resistance is small (low-voltage ranges) than when the multiplier resistance is large (high-voltage ranges). This condition is known as the *swamping effect*.

The ac scales of a meter are never more accurate than the dc scales, usually, they are *less* accurate. This is because of the inaccuracy of the dc circuit (multiplier, movement, and scales). Accuracies for a typical meter are ±3 percent of full scale for direct current and ±5 percent for alternating current. An electronic volt-ohmmeter will sometimes have the same accuracy for both ac and dc scales (typically ± 3 percent).

The effects of *frequency* must be considered in determining the accuracy of ac scales. A typical VOM provides accurate ac voltage indications from 15 hertz (Hz) up to 10 kHz, possibly up to 15 or 20 kHz, but rarely beyond that frequency. This means that VOM reading in the high audio range may be inaccurate.

It should be noted that an ac meter may provide readings well beyond its maximum rated frequency, but these readings will not necessarily be accurate. A typical electronic volt-ohmmeter provides accurate ac voltage indications from 15 Hz up to about 3 megahertz (MHz). Of course, the frequency range of a meter can be extented by use of an RF probe. However, the probe and meter must be calibrated together, as described in Sec. 2-7.

As shown in Fig. 2-18, many electronic volt-ohmmeters permit ac voltages to be read as an rms value or a peak-to-peak value, whichever is convenient. Note that the meter circuit responds to the *average* value in either case but that the scales provide for rms or peak-to-peak indications.

Note that the rms scales are accurate only for a *pure sine wave*. If there is any distortion, or if the voltage contains any component other than a pure sine wave, the readings will be in error. On the other hand, peak-to-peak readings will be accurate on any type of wave form, including sine waves. Both rms and peak voltages are found in troubleshooting service literature.

Ac voltage measurements are usually made with a *blocking capacitor* in series with one of the test leads. The capacitor is part of the meter circuits. On a typical VOM, the test lead is connected to a terminal marked OUTPUT or some similar function. This blocks any direct current present in the circuit being measured from passing to the meter circuit. Such direct current may or may not damage the meter, depending upon circuit conditions.

Ac voltage can also be measured on most meters without the blocking capacitor. On a typical VOM, this is done by connecting the test leads to the same terminals used for dc voltage measurements (usually marked PLUS and MINUS or COMMON and PLUS).

Some older meters have *half-wave rectifier* circuits or use a *half-wave probe* circuit. On such meters, the ac voltage readings may be affected by a condition known as *turnover*, which occurs when there are *even harmonics* present in the voltage being measured. Turnover will show up at different ac voltage readings when the test leads are reversed. Turnover should not occur when odd harmonics are present, when there are no harmonics, or when a full-wave rectifer (or full-wave probe) is used.

dB scales. Most VOMs are provided with decibel (dB) scales. In troubleshooting, about the only practical use for dB scales is to measure the power gain of amplifier stages or circuits. Because most service literature shows gain in terms of voltage rather than power, the dB scales are not often used in troubleshooting.

When measuring decibels, the ac voltage circuits of the meter are used in the normal manner except that the readout is made on the dB scales. Inexperienced troubleshooters are often confused by the dB scales. The following discussion should clarify their use.

The dB scales represent *power* ratios, not voltage ratios. In most cases, 0 dB is considered the power of 1 milliwatt (mW), which is 0.001 watt (W), across a 600 Ω pure resistive load. This also represents 0.775 V rms across a 600 Ω pure resistive load. The term *decibel meter* (dBM) is sometimes used to indicate this system (1 mW across 600 Ω).

The dB scale is related directly to only *one* of the ac scales, usually the lowest scale. The VOM range selector must be set to that ac scale if readings are to be taken directly from the dB scale. If another ac scale is selected by the range selector, a certain dB value must be added to the indicated value.

For example, on the VOM shown in Fig. 2-17, the dB scale is related directly to the 3 V ac scale. If the range selector is set to 3 V ac, the dB scale be read out directly. (Note that the 0 dB mark is aligned with the 0.775 V point on the 3 V ac scale.) If the range selector is set to 8, 40, or 160 V ac, it is necessary to add 8.5, 22.5, or 34.5 dB to the indicated dB scale reading. These values are printed on the meter face (lower right-hand corner) and are applicable to that meter only. Always consult the meter face (or service literature) for data regarding the dB scales.

Note that the dB scale readings will not be accurate if

1. voltages are other than pure sine waves

2. load impedances are other than pure resistive

3. load is other than 600 Ω

If the load is other than 600 Ω, it is possible to apply a correction factor. The decibel is based on this mathematical function:

$$dB = 10 \log \frac{\text{power output}}{\text{power input}}$$

The power will change by the corresponding ratio when resistance is changed (power increases if resistance decreases and voltage remains the same). Therefore, it is possible to convert the function to $10 \log R_2/R_1$, where R_2 is 600 Ω and R_1 is the resistance value of the load.

For example, assume that the load resistance is 500 Ω instead of 600 Ω and that a 0 dB indication is obtained (0.775 V rms).

$$\frac{600}{500} = 1.2$$

The log of 1.2 is 0.792.

$$10 \times 0.0792 = 0.792$$

Thus, 0.792 (0.8 for practical purposes) must be added to the 0 dB value to give a true reading of 0.8 dB.

Using dB scales. The following discussion should clarify use of dB scales in measuring the input-output relationships of a particular circuit during troubleshooting.

If the *input* and *output load impedances* are 600 Ω (or whatever value is used on the meter scale), no problem should be found. Simply make a dB reading at the input and the output (under identical conditions), subtract the smaller dB reading from the larger, and note the dB gain (or loss).

For example, assume that an amplifier is being monitored during troubleshooting and that the input shows 3 dB with 13 dB at the output. This represents a 10 dB gain. If the output is 3 dB with 13 dB input, this represents a 10 dB loss. Remember that these are *power* gains and losses.

If the *input* and *output load impedances* are not 600 Ω but are *equal* the relative dB gain or loss is correct, even though the absolute dB reading is incorrect.

For example, assume that the input and output load impedances of the amplifier being serviced are 50 Ω and that the input shows 3 dB with 13 dB at the output. There is still a 10 dB *difference* between the two readings, and there is a power gain of 10.

If the *input* and *output load impedances* are *not equal*, the relative dB gain or loss indicated by the meter scales will be incorrect.

For example, assume that the input impedance is 300 Ω, the output impedance is 8 Ω, the input shows +7 dB, and the output shows +3 dB on the scales of a meter using a 600 Ω reference. There is an apparent loss of 4 dB (7 dB input − 3 dB output). However, a 300 Ω input measured on a 600 Ω meter requires a correction of about +3 dB (600/300 = 2, 10 log 2 = 3). Thus, the actual input is about 10 dB (7 + correction of 3). An 8 Ω output measured on a 600 Ω meter requires a correction of about +19 (600/8 = 75; 10 log 75 = 19). Thus, the actual output is about 22 dB (3 + correction of 19). With an input of 10 dB and an output of 22 dB, there is an *actual* power gain of 12 dB (22 − 10 = 12).

2-2.6 Parallax Problems

There are many factors that can affect the accuracy of meter readings. Some of these have to do with the operator; others are dependent upon the meter. A complete discussion of meter accuracy is beyond the scope of this book. However, the problem of parallax is common to all readings made during troubleshooting with all types of meters (except digital meters).

In nondigital meters, parallax is an error in observation that occurs when the operator's eye is not directly over the meter pointer, as shown in Fig. 2-20. This will cause the reading to appear at the right or left of the actual indication, resulting in an erroneous high or low reading. Some meter manufactures minimize this problem by placing a mirror behind the pointer on the scale, as shown in Fig. 2-21.

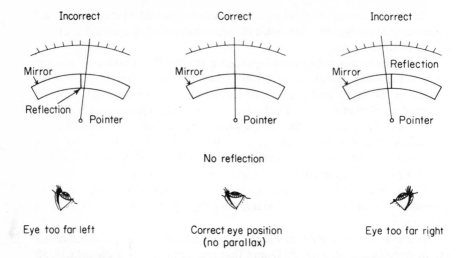

FIGURE 2-20   Using antiparallax mirrored scales (B & K Manufacturing)

FIGURE 2-21   Simpson Instruments Model 270 (with mirrored scale)

To use a mirrored scale most effectively, you should close one eye; the pointer and its reflection will appear to coincide, as shown in Fig. 2-20.

2-2.7 Basic Meter Operating Procedures

This section describes the basic operating procedures for a VOM or an electronic volt-ohmmeter (resistance, voltage, current, and dB measurements).

Meter operating precautions. In addition to the general troubleshooting safety precautions described in Sec. 2-1, the following specific precautions should be observed when operating any type of meter.

1. Even if you have had considerable experience with meters, always study the service literature of any meter with which you are not familiar.

2. *Never* measure a voltage with the meter set to measure current or resistance. To do so will damage the meter movement. Similarly, never measure a current with the meter set to measure resistance.

3. Always *start* voltage and current measurements on the *highest* voltage or current scale. Then, switch to a lower range as necessary to obtain a good center-scale reading.

4. Do not attempt to measure ac voltages or currents with the meter set to measure direct current. This could damage the meter movement and will produce error in the meter readings. Usually, no damage will result (but consult the service literature) if dc voltages or currents are measured with the meter set to measure alternating current. However, the readings will be in error.

5. Use only shielded probes. Never allow your fingers to slip down to the metal probe tip when the probe is in contact with a hot circuit.

6. Avoid operating a meter in strong magnetic fields. These fields (such as those produced by the degaussing coil used in color television service) can cause inaccuracy in the meter movement and can damage it. Most better-quality meters are well shielded against magnetic interference. However, the meter face is still exposed and is subject to the effects of magnetic fields.

7. Most meters have some maximum input voltage and current specified in the service literature (or on the meter scales). Do not exceed this maximum. Also, do not exceed the maximum line voltage or use a different power frequency on those meters that operate from line power.

8. Avoid vibration and rough treatment. Like most electronic equipment, a meter is a delicate instrument.

9. Do not attempt repair of a meter unless you are a *qualified* instrument technician. If you must adjust any internal controls on a meter, follow the instructions in the service literature.

10. Study the circuit being serviced before making any test connections. Try to match the capabilities of the meter to the circuit. For example, if the circuit has a range of measurements to be made (alternating current, direct current, radio frequency, modulated signals, pulses, or complex waves), it may be necessary to use more than one instrument. Most meters will measure direct current and low-frequency alternating current. If an unmodulated RF carrier is to be measured, use an RF probe. If the carrier to be measured is modulated with low-frequency signals, use a demodulator probe. If pulses, square waves, or complex waves (combinations of alternating current, direct current, and pulses) are to be measured, meters with peak-to-peak readings will provide the only meaningful indications.

11. Remember that all voltage measurements are made with the meter in *parallel* or across the circuit and that all current measurements are made with the meter in *series* with the circuit.

2-2.8 Basic Ohmmeter (Resistance) Measurements

The first step in making a resistance measurement is to zero the meter on the resistance range that is to be read. The meter can be zeroed on other ranges. Some meters will remain constant for all ranges; on other meters, the ohmmeter zero will change for each range.

The meter is usually zeroed by touching the two test prods together and adjusting the ZERO-OHMS or OHMS control until the pointer is at zero. This is usually at the right end of the scale for a VOM and at the left end for an electronic volt-ohmmeter, as shown in Fig. 2-22(a).

Once the ohmmeter is zeroed, connect the test prods across the resistance to be measured, as shown in Fig. 2-22(b). Read the resistance from the ohmmeter scale. Make certain to apply any multiplication indicated by the range-selector switch.

For example, if an indication of 3 is obtained with the range selector at $R \times 10$, the resistance is 30 Ω. It should be possible to set the range selector to $R \times 1$ and obtain a direct reading of 30 Ω. However, it may or may not be necessary to zero the ohmmeter when changing ranges.

Two major problems must be considered in making any ohmmeter measurements.

Power removal. There must be *no* power applied to the circuit being measured. Any power in the circuit can damage the meter and cause an incorrect reading. Remember, capacitors often retain their charge *after* power is turned off. With power off, short across the circuits to be measured

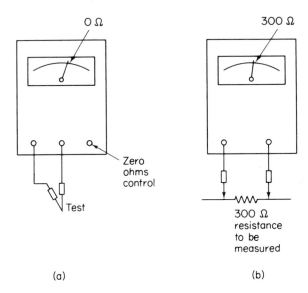

(a)   (b)

FIGURE 2-22   Basic resistance-measurement procedure

(with a screwdriver or similar tool) in order to discharge any capacitance. Then, make the resistance measurement.

Parallel resistances. Make cerain that the circuit or component to be measured is not in parallel with (or shunted by) another circuit or component that will pass direct current. For example, assume that the value of resistor R_1 shown in Fig. 2-23 is to be measured. If the battery in the ohmmeter is connected so that the diode CR_1 is forward-biased, current can flow through the transformer T_1 winding, diode CR_1, and choke coil L_1. All these components have some dc resistance, the total of which is in parallel with resistor R_1.

The simplest method to eliminate the parallel resistance is to disconnect one lead of the resistance. This technique is shown in Fig. 2-23(b).

2-2.9 Basic Voltmeter (Voltage) Measurements

The first step in making a voltage measurement is to set the range. Always use a range that is *higher* than the anticipated voltage. If the approximate voltage is not known, use the highest voltage range of the meter initially, and then select a range that will allow you to obtain a good mid-scale reading.

Next, set the function selector to alternating current or direct current, as required. In the case of direct current, it may also be necessary to select either plus ($+$) or minus ($-$) by means of the function switch. On simple meters, polarity is changed by switching the test leads.

On an electronic voltmeter, the next step is to zero the meter. This should be done *after* the range and function have been selected. Touch the

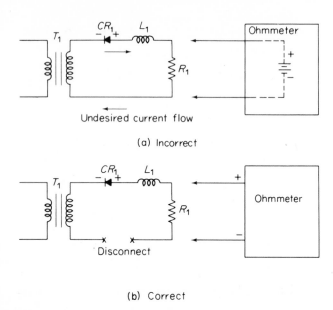

(a) Incorrect

(b) Correct

FIGURE 2-23   Avoiding errors in resistance measurements because of parallel resistance

test leads together, and adjust the ZERO control for a zero indication on the voltage scale to be used, as shown in Fig. 2-24(a).

Remember that *all* voltage measurements (alternating current, direct current, plus, minus, and decibels) are made with the meter in *parallel* across the circuit and voltage source, as shown in Fig. 2-24(b). This means that some of the current normally passing through the circuit being tested will be passed through the meter. In the case of a VOM, where the total meter resistance (or impedance) is low, considerable current may pass through the meter. This may or may not affect circuit operation.

For example, an oscillator that develops a small voltage across a high circuit impedance can be prevented from oscillating if a VOM is used to measure the voltage. This is shown in Fig. 2-24(c), in which a voltmeter with 1,000 Ω resistance is used to measure an oscillator output of 100 mV (developed across an impedance of 100,000 Ω). The current will divide itself across the two circuits at a ratio of about 99:1, with most of the current going through the meter. The same voltage is fed back to the oscillator circuit to sustain oscillation. If most of the current is passed through the meter, the feedback will not be sufficient to produce any oscillation.

The problem of parallel current drain does not occur in an electronic meter except when a voltage is measured across a very high impedance circuit. A typical electronic meter has an input impedance of 10 to 15 megohms (MΩ), with FET meters having an input of about 100 MΩ. If the circuit imped-

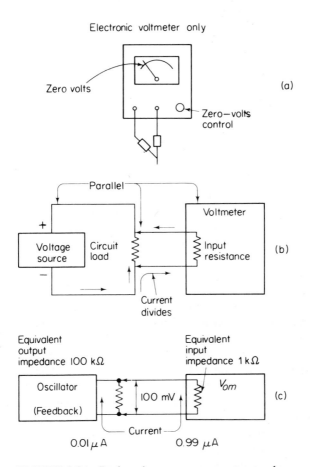

Electronic voltmeter only

Zero volts

Zero—volts control

(a)

Parallel

Voltmeter

+

Voltage source

Circuit load

Input resistance

(b)

−

Current divides

Equivalent output impedance 100 kΩ

Equivalent input impedance 1 kΩ

Oscillator

(Feedback)

100 mV

V_{om}

(c)

Current

0.01 μA

0.99 μA

FIGURE 2-24   Basic voltage-measurement procedure

ance is near this value, the current will divide itself between the circuit and the meter, possibly resulting in an erroneous reading.

Measuring alternating current in the presence of direct current. If you want to measure *alternating current only* but direct current is also present in the circuit, the OUTPUT or AC ONLY function can be selected, thereby switching a capacitor into the meter input. On a VOM, this is done by connecting the free test lead to the OUTPUT terminal. On an electronic meter, alternating current is often selected by means of a switch on the probe. On some meters, alternating current is always measured with a coupling capacitor at the input. In any event, the direct current is blocked, and the alternating current is passed.

Measuring direct current in the presence of alternating current. If you want to measure *direct current only*, but alternating current is also present in the circuit, there are several possible solutions. If the alternating current is of high frequency, it is possible that the meter movement will not respond and that there will be no ac indications when the meter is set to measure direct current. If the ac voltage is low in relation to the direct current being measured, it is also possible that the meter will not be affected.

If the meter is affected by the presence of alternating current, one solution is to connect a capacitor across the test leads. This will provide a bypass for the alternating current but will not affect the direct current. However, the capacitor may affect the operation of the circuit. Also, remember that the capacitor will be charged to the full value of the direct current.

In some cases, it is possible to use a high-voltage or attenuator probe to measure direct current in the presence of alternating current. The series resistance of the probe, combined with the natural capacitance between the probe's inner and outer conductors, or shields, forms a low-pass filter. This filter will have no effect on direct current but will reject alternating current.

The fact that electronic voltmeters usually use some form of probe makes these instruments better suited than VOMs to measure direct current in the presence of alternating current.

2-2.10 Basic Ammeter (Current) Measurements

The first step in making a current measurement is to set the range. Always use a range that is higher than the anticipated current. If the approximate current is not known, use the highest current range first, and then select a lower range so that a good mid-scale reading can be obtained.

(Note that most electronic meters do not have a provision for measuring current, primarily because of their high input impedance. Because current must pass through the meter's input circuit, there is a voltage drop across the meter. In an electronic meter, the voltage drop could be very high. In some electronic meters, current is measured by connecting directly to the meter and shunts, thus bypassing the high input impedance.)

In many meters, selecting a current range involves more than positioning a switch. A typical VOM requires that the test leads be connected to different terminals. For example, the high-current range could require that the test leads be connected to the −10 A and +10 A terminals. On the low-current range, the COMMON and 50 μA terminals could be used. On all other current ranges, the COMMON and PLUS terminals can be used. No matter what terminal arrangement is used, the range selector must be set to the appropriate range in all cases.

Once the current range has been selected, set the function selector to alternating current or direct current, as required. (Many VOMs will not measure alternating current; therefore, either plus or minus direct current must be selected.)

Note that when the *lowest* current scale is selected, such as 50 μA, the meter is actually functioning as a *voltmeter*. The meter movement is placed (without a shunt) in series with the circuit. Thus, any sudden surges of current can damage the meter movement. This is a problem especially when both alternating current and direct current are present in the circuit being measured. If the alternating current is of higher frequency, it will probably have little effect on the meter movement. Lower-frequency alternating current can combine with the direct current and possibly cause errors or meter-movement burnout.

All current measurements are made with the meter in *series* with the circuit and power source, as shown in Fig. 2-25. This means that all the current normally passing through the circuit under test will be passed through the meter. This may or may not affect circuit operation.

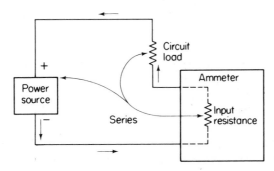

FIGURE 2-25 Basic current-measurement procedure (measurements made with meter in series with circuit and power source)

2-2.11 Basic dB Measurements

The procedures for measurement in decibels is similar to that for ac voltage measurement, with two exceptions:

1. The OUTPUT function is always used for dB measurements.

2. The dB scales are used instead of the ac rms or peak-to-peak scales.

When making dB measurements, use the basic voltage-measurement procedures outlined in Sec. 2-2.9, and observe the precautions concerning dB scales described in Sec. 2-2.5.

2-3. DIGITAL METERS

To understand the operation of digital meters (or any digital instrument), it is necessary to have a full understanding of digital *logic* circuits. Most digital equipment (meters, counters, computers, and so on) is made up of logic building blocks (gates, registers, and so on) that are interconnected to perform various functions (mathematical operation, conversion, readout, and so forth). A detailed discussion of these circuits is beyond the scope of this book. Instead, we will concentrate on the basic readout functions of digital meters. From a troubleshooting standpoint, this is the most valuable approach because you will generally be more interested in interpreting meter readout than in understanding overall meter function.

A knowledge of electronic counters is particularly necessary to a full understanding of digital meters because a digital meter performs two basic functions: conversion of voltage (or some other quantity being measured) to time or frequency and conversion of the time or frequency data to a digital readout. In effect, a digital meter is a conversion circuit (voltage to time and so on) plus an electronic counter for readout.

Basic digital meter. There are digital instruments to measure ac and dc voltages, direct currents, resistance, and other variables. The most popular digital meter is the digital voltmeter (DVM). Such instruments display measurements as *discrete numerals*, rather than as the pointer deflection on a continuous scale commonly used in analog meters. Direct numerical readout reduces human error, eliminates parallax error, and increases reading speed. On some digital meters, automatic polarity- and range-changing features reduce operator training, measurement error, and possible instrument damage as a result of overload.

Figure 2-26 shows the operating controls and readout of a typical DVM. Note the simplicity of the controls. Once power is turned on, the operator has only to select the desired range and connect the meter to the circuit. The readout is automatic. On some digital meters, the range is changed automatically, further simplifying operation.

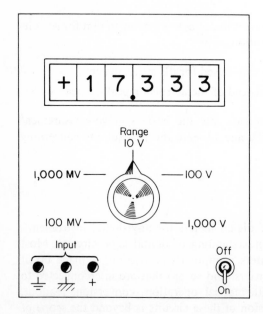

FIGURE 2-26 Typical digital voltmeter panel controls and indicators

Because of this simplicity of operation, no special precautions need to be observed in the use of DVMs in troubleshooting, although, of course, it is necessary to follow all precautions described in the meter's service literature (or operating manual). Similarly, all the general operating precautions described for VOMs and electronic volt-ohmmeters apply to DVMs use in troubleshooting.

2-4. BRIDGE-TYPE TEST EQUIPMENT

Many quantities in electronics (such as capacitance, impedance, admittance, inductance, and conductance) are measured by means of bridge circuits and bridge-type test equipment. These instruments operate on the balance or null principle or on the principle of comparison against a standard. In general, bridge-type instruments are used for precision laboratory measurements, rather than for troubleshooting. But the bridge circuits used in *capacitor checkers* are an exception. As discussed throughout this book, capacitors can be checked by means of voltmeters and ohmmeters, both in circuit and out of circuit. However, in some circumstances, it is convenient to test a capacitor for proper value, as well as for leakage, shorts, breakdown, and the like. A capacitor checker will provide these functions and is thus considered a troubleshooting test instrument by some technicians.

Wheatstone bridge. The bridge circuits found in capacitor checkers are a form of Wheatstone bridge that measures an unknown resistance R_x in terms of calibrated standards of resistance, as shown in Fig. 2-27. When

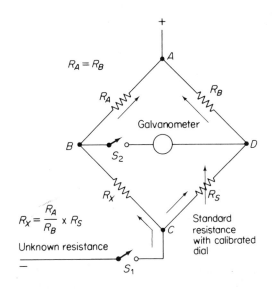

FIGURE 2-27   Basic Wheatstone resistance bridge

switch S_1 is closed, current flows in the direction indicated by the arrows, and there is a voltage drop across all four resistors. The drop across R_A is equal to the drop across R_B (provided that R_A and R_B are of equal resistance value). Variable resistance R_S is adjusted so that the galvanometer reads zero (center scale) when switch S_2 is closed. At this adjustment, R_S is equal to R_X in resistance. By reading the resistance of R_S (from a calibrated dial), the resistance of R_X is determined.

Capacitance bridges. The capacitance bridge is similar to the basic Wheatstone bridge except that an ac power source is required because a capacitor will not pass direct current, as shown in Fig. 2-28. Variable capacitance C_N is adjusted so that the detector meter reads zero. At this setting, C_N is equal to C_X (the capacitance being tested). The capacitance value of C_X is then equal to that indicated on the calibrated dial attached to C_N.

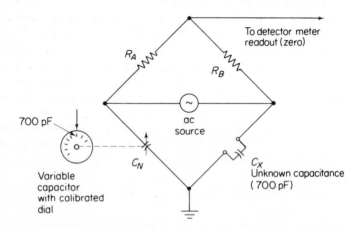

FIGURE 2-28   Basic standard capacitance circuit used in capacitance checkers

2-5. SIGNAL GENERATORS

Next to a meter, a signal generator is the most useful tool in troubleshooting. Without some form of signal generator, you are entirely dependent upon signals broadcast from transmitting stations or available from equipment containing oscillators or pulse sources. Under these circumstances, you will have no control over the frequency, amplitude, or modulation of such signals.

But with a signal generator of appropriate type, you can duplicate transmitted signals or produce special signals (such as digital pulses) required for troubleshooting a particular piece of equipment. Similarly, the frequency, amplitude, and modulation characteristics of these signals can be controlled.

2-5.1 Signal Generator Basics

An oscillator (audio, radio frequency, pulse, and so on) is the simplest form of signal generator. At the most elementary level of troubleshooting, a single-stage audio or RF oscillator can serve the purpose of providing a signal source. An example of this is the simple *probe-type* oscillator used to trace signals through an audio amplifier or broadcast receiver.

Except in basic troubleshooting situations, such an oscillator (sometimes known as a *pencil-type noise generator*) has many obvious drawbacks. For example, to check the selectivity of a receiver or some other instrument, the signal source must be *variable in amplitude*. To check the detector or audio portions of receivers, the oscillator should be capable of *internal* and/or *external modulation*. These characteristics are not available in the pencil-type unit, which is essentially a solid-state pulse generator with a fast-rise waveform output and no adjustments. As a result, even the least expensive shop-type (or even kit-type) generators have many advantages over the basic oscillator or generator circuit.

2-5.2 RF Signal Generators

RF signal generators are used in both commercial and laboratory troubleshooting. There are no basic differences between the commercial and laboratory generators; that is, both instruments will produce RF signals capable of being varied in frequency and amplitude and capable of internal or external modulation. However, the laboratory instruments incorporate several refinements not found in commercial equipment, as well as a number of quality features. (This accounts for the wide difference in price.) The following paragraphs summarize the differences between commercial and laboratory RF generators.

Output meter. In most commercial generators, the amplitude of the RF output is either unknown or approximated by means of dial markings. The laboratory generator incorporates an output meter. This meter is usually calibrated in microvolts so that the actual RF output can be read directly.

Percentage-of-modulation meter. Most commercial generators have a fixed percentage of modulation (usually about 30 percent). Laboratory generators provide for a variable percentage of modulation and a meter to indicate this percentage. Some generators have two meters (one for output amplitude and one for modulation percentage). Some generators use the same meter for both functions.

Output uniformity. Commercial generators vary in output amplitude from band to band but (usually) cover their range by means of harmonics or beat notes. Laboratory generators have a more uniform output over their entire operating range and cover the range with pure fundamental signals.

Wideband modulation. Generally, the oscillator of a commercial generator is modulated directly. This can result in undesired frequency modulation, just as it does in the case of a transmitter using a directly modulated oscillator. The oscillator of a laboratory generator is never modulated directly (unless it is designed to produce an FM output). Instead, the oscillator is fed to a wideband amplifier, where the modulation is introduced. Thus, the oscillator is isolated from the modulating signal.

Frequency drift. Because a signal generator must provide continuous tuning across a given range, some type of variable-frequency oscillator (VFO) must be used. As a result, the output is subject to frequency drift, instability, modulation (by noise, mechanical shock, power supply ripple), and other problems associated with VFOs. Frequency instability does not present too great a problem in commercial work. Also, it is possible to calibrate a signal generator at or near the most used frequency points. However, laboratory signal generators must provide a *known degree* of frequency accuracy over their entire operating range.

In laboratory generators, the output is less subject to frequency drift because of the incorporation of temperature-compensating capacitors. Similarly, the effects of line-voltage variations are offset by regulated power supplies. The better generators also have more elaborate shielding, especially for the output-attenuator circuits, where radio frequency is most likely to leak. The leakage of radio frequency from signal generators is something of a problem in many laboratory applications (such as receiver-sensitivity tests).

Band spread. Commercial generators usually have a minimum number of bands for a given frequency range. This makes the tuning-dial or frequency-control adjustments more critical as well as difficult to see. Laboratory generators usually have a much greater band spread; that is, they cover a smaller part of the frequency range in each band.

Typical RF generator circuits. Figure 2-29 shows the basic circuit of a commercial RF generator. Figure 2-30 shows the corresponding circuit for a laboratory RF generator.

In the commercial generator (Fig. 2-29), the RF oscillator is selected by changing the coil of the tuned tank circuit (one coil for each band). The actual frequency (within the band) is controlled by the setting of the tuned tank capacitor (usually geared or directly coupled to a frequency dial). The amplitude of the RF output is controlled by coarse and fine attenuators. The audio-oscillator frequency is usually fixed at 60, 400, or 1,000 Hz. This audio output is available at the front panel or can be used to modulate the RF oscillator. An external modulating signal can also be used to modulate the RF oscillator.

In the laboratory generator (Fig. 2-30), the RF oscillator is isolated from the modulating signal by a modulator stage, usually a wideband amplifier. The RF oscillator may also be isolated from the output by an emitter follower or some similar buffer stage. The meter circuit monitors either the RF

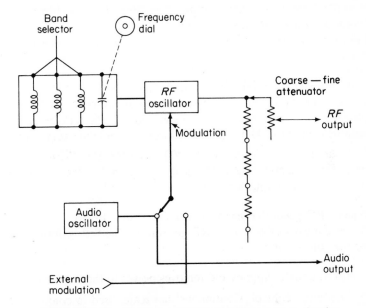

FIGURE 2-29   Typical commercial RF generator circuit

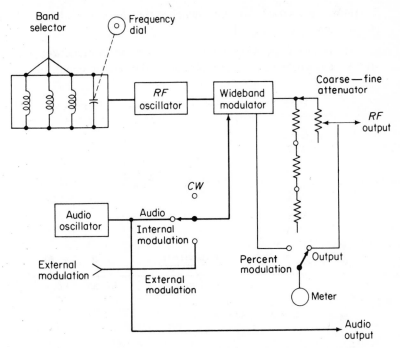

FIGURE 2-30   Typical laboratory RF generator circuit

output or the percentage of modulation, depending on the position of a front-panel control. Some generators use two meters for this purpose.

Typical RF generator outputs. The outputs of a typical RF generator are as follows:

1. An RF signal, variable in frequency from about 100 kHz to 250 MHz (up to 1,000 MHz for some laboratory generators). Usually, this frequency range is covered in four to eight bands. The RF signal output is variable from 0 to 100,000 μV (up to 1 or 2 V for laboratory generators).

2. An audio signal, usually 60, 400, or 1,000 Hz. With most signal generators, this audio signal can be used to modulate the RF output or is available as an output, or both.

Typical RF generator controls and indicators. Figure 2-31 shows the front-panel controls of a typical RF generator. The following are descriptions of the control functions:

1. POWER switch. Applies and removes power to the instrument.

2. FREQUENCY control. Continuous dial scale, used in conjunction with the RANGE switch to select output frequency.

3. STEP ATTENUATORS. Six slide switches that have a total capability of 96 dB attenuation on the RF output.

4. FINE ATTENUATOR. Continuous dial, used in conjunction with the STEP ATTENUATORS to adjust amplitude of the output signal.

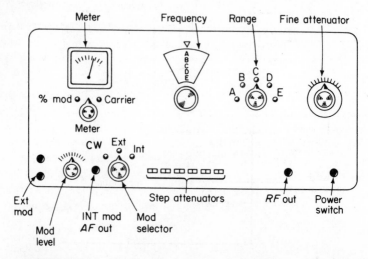

FIGURE 2-31 Typical RF generator controls and indicators

5. RF OUT. The modulated or unmodulated RF output from the generator is available at the connector. The output connector accommodates a coaxial cable (usually supplied with the instrument). This cable should always be used during troubleshooting.

6. EXT MOD. The external modulation signal to be superimposed on the RF output is fed into these terminals. The MOD SELECTOR switch must be set to EXT MOD for these terminals to be placed in operation.

7. INT MOD/AF OUT. The internal modulation signal (usually 60 or 400 Hz) is available at this connector when the MOD SELECTOR switch is set to INT.

8. MOD SELECTOR. In the INT MOD position, the generator output is modulated by the 60 or 400 Hz internal signal. Simultaneously, the modulating signal is available at the INT MOD/AF OUT connector. In the EXT MOD position, the generator output is modulated by external signals applied to the EXT MOD terminals. In the continuous wave (CW) position, all modulation is removed.

9. METER. Indicates the carrier's output level (in microvolts) or percentage of modulation, depending upon the position of the METER switch.

10. MOD LEVEL. Sets level or percentage of modulation.

Typical RF generator operating procedures. The following steps describe the basic operating procedures for a typical RF signal generator.

1. Observe the general safety precautions discussed in Sec. 2-1.

2. Set the POWER switch to OFF. Connect the power cord to an ac outlet.

3. Set the POWER switch to ON, and allow 15 minutes (or whatever time is recommended in the service literature) for warm-up. All but the very best signal generators will have some tendency to drift in frequency as they warm up.

4. Connect the test cable to the RF OUT connector. Most generators are supplied with a cable for the RF output. These cables are specially designed for use with the instrument and should always be used in preference to other types of cables. Such cables are shielded throughout their lengths to prevent excessive radiation of the output signal and to minimize hum pickup. The cables are usually terminated in a 50 or 75 Ω impedance and are usually unbalanced. Some signal generators designed specifically for television service have a balanced 300 Ω output.

5. Set the RANGE switch to the appropriate frequency range or band.

6. Set the FREQUENCY dial to the exact frequency. Make sure that you read the frequency on the *correct band*, as selected by the RANGE control.

7. Set the MOD SELECTOR switch to INT MOD (for internal modulation), CW (for no modulation), or EXT MOD (for external modulation), as desired.

8. Set the STEP ATTENUATORS (coarse attenuation) and FINE ATTENUATOR control, as necessary. Set the METER switch to CARRIER; then, adjust the attenuators for the desired RF output amplitude (usually specified in microvolts).

If no signal-amplitude value is specified for a particular trouble-shooting procedure, use the following:

Set both attenuator controls to give the *smallest* output necessary to obtain a waveform of the desired amplitude on an oscilloscope (or a mid-scale indication on a meter). If too strong a signal is injected into the circuit being tested, it is possible that overloading may cause distortion. This can lead to false conclusions during troubleshooting.

The vertical gain of the oscilloscope (Sec. 2-6) should be set at or near the *maximum* gain point so that the oscilloscope furnishes a good share of the signal amplification. If you use a meter to monitor the circuit you are troubleshooting, use the *lowest* scale that will provide a good *mid-scale* reading.

9. Set the METER switch to % MOD. Adjust the MOD LEVEL control for the desired percentage of modulation.

10. Clip the *ground lead* of the RF output cable to the *ground* of the circuit you are troubleshooting. Connect the probe or clip of the RF output cable to the point of *signal injection*.

11. In some signal generators, the internal circuits are connected directly to the output terminals. A connection between the output terminals or cable and an external circuit carrying dc voltage may result in damage to the signal generator. If it becomes necessary to connect a generator that does not have an internal blocking capacitor to a power circuit, an external blocking capacitor of suitable value should be used *in series with the output terminal*.

2-5.3 Audio Generators and Function Generators

Audio generators (also known as *audio oscillators*) are particularly useful in troubleshooting all types of audio amplifiers, as well as the audio circuits of other equipment (such as television and AM-FM receivers). Audio generators can also be used as modulation sources for RF signal generators. As is the case with RF signal generators, audio generators in their simplest form are essentially audio oscillators. For troubleshooting purposes, the audio output is tunable in frequency over the entire audio range (and beyond) and is variable in amplitude.

Early audio generators produced only sine waves. However, for use in present-day troubleshooting techniques, most commercial audio generators also produce *square waves* at *audio frequencies*. Some laboratory audio generators are referred to as *function generators* because they produce various functions: sine waves, square waves, triangular, and/or sawtooth waves. All these functions are produced at audio frequencies.

The major differences in audio generators are quality differences rather than special features. For example, laboratory audio generators are less subject to frequency drift and line-voltage variations. The effects of hum or other line noises are minimized by extensive filtering. Accuracy and dial resolution are generally better for laboratory generators. This makes the tuning-dial or frequency-control adjustments less critical. Laboratory generators also have a more uniform output over their entire operating range. Commercial audio generators may vary in amplitude from band to band.

Typical audio-generator circuits. Figure 2-32 shows the block diagram of a typical commercial audio generator. Operation of the circuit is as follows:

The basic oscillator circuit is composed of transistors Q_1 and Q_2. The oscillator frequency is determined by the resistance-capacitance (RC) time constants in a feedback circuit. The frequency range is selected by changing

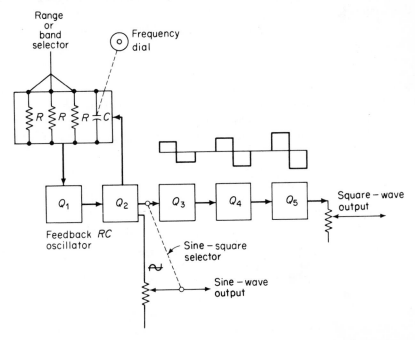

FIGURE 2-32   Typical commercial audio-generator circuit

resistance values (the R of the RC circuit); tuning within a given range is controlled by a variable capacitance (the C of the RC circuit).

With the SINE-SQUARE selector in the SINE position, the output is taken from the emitter of Q_2 and fed through the output-divider network to the output jack. With the SINE-SQUARE selector in the SQUARE position, the oscillator is taken from the collector of Q_2 and fed to Q_3, where the square-wave–shaping action begins. The sine wave is reshaped to a square wave by the amplification and limiting action of Q_3, Q_4, and Q_5. The square-wave signal is taken from the emitter of Q_5 and fed through the output-divider network to the output jack.

Figure 2-33 shows the basic circuit of a laboratory audio generator. There are a number of audio-generator circuits that qualify as laboratory troubleshooting equipment and provide the high-frequency stability and low-distortion characteristics. However, the Wien bridge RC oscillator has become the standard.

As shown in Fig. 2-33, the basic circuit is a two-stage amplifier with both negative and positive feedback. Positive feedback for sustaining oscillations is applied through the frequency-selective network R_1C_1–R_2C_2 of the Wien bridge. The amplitude response is maximum at the same frequency at which the phase shift through the network is zero. Oscillations are thus sustained at this frequency.

The resonant frequency F_0 is expressed by the equation

$$F_0 = \frac{1}{6.28\ RC}$$

where $R_1 = R_2$ and $C_1 = C_2$.

Unlike inductance-capacitance (LC) circuits used in RF generators, where the frequency varies inversely with the square root of C, the frequency of the RC Wien bridge oscillator varies inversely with C. Thus, frequency

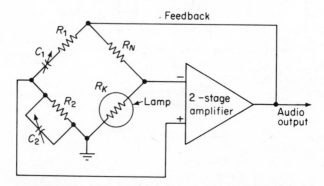

FIGURE 2-33 Typical laboratory audio-generator circuit

variations greater than 10 to 1 are possible with a single sweep of an air-dielectric tuning capacitor. Range switching is usually done by switching the resistors.

The negative-feedback loop involves the other pair of bridge arms, R_N and R_K. In a Wien bridge oscillator, R_K is some form of temperature-sensitive resistor with a positive temperature coefficient. Generally, R_K is an incandescent lamp operated at a temperature level lower than its illumination level. The lamp is sensitive to the amplitude of the driving signals and therefore adjusts the voltage-division ratio of the branch accordingly. Thus, as the amplitude of oscillation increases, the resistance of R_K increases, reducing the gain of the amplifier and restoring the amplitude to normal.

Solid-state audio generators. A different type of amplitude stabilization is used in solid-state oscillators. Because the current drawn by a lamp is incompatible for use with transistors and low-voltage power sources, solid-state generators use an output- or peak-detector circuit that provides a bias voltage proportional to the oscillator's output voltage.

Such a circuit is shown in Fig. 2-34. As the amplitude of the amplifier changes, the detector circuit sends an error signal to the automatic gain control, which contains an FET. The purpose of the AGC circuit is to control the oscillator gain continuously to maintain a loop gain of 1. The resistance of the AGC circuit can be varied slightly to change the divider ratio of the negative-feedback network. An error in output voltage is detected by the output detector and sent to the AGC circuit. This changes the resistance ratio in the negative-feedback loop, thus bringing the output back to a constant level.

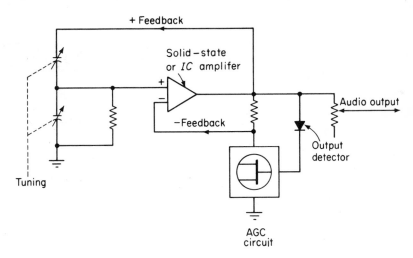

FIGURE 2-34   Typical solid-state audio-generator circuit

Typical audio-generator outputs. The outputs of a typical audio generator are as follows:

A sine wave, variable in frequency from about 20 Hz to 200 Hz, or possibly as high as 1 MHz. Usually, this frequency range is covered in several bands. The sine-wave output is variable from 0 to about 8 or 10 V.

A square wave, variable in frequency and amplitude over the same range as the sine wave.

Typical audio-generator controls and indicators. Figure 2-35 shows the front-panel controls of a typical audio generator. The following are descriptions of the control functions:

1. FREQ RANGE. Selects the multiple to be applied to the frequency setting of the TUNING CONTROL dial.

2. SINE-SQUARE ATTENUATOR. Selects either sine-wave or square-wave output and the output-voltage range.

3. TUNING CONTROL. Continuous, single-scale dial, used in conjunction with the FREQ RANGE switch to select the output frequency.

4. OUTPUT. Adjusts the amplitude of the output signal within range of SINE-SQUARE ATTENUATOR.

5. OFF-ON/LINE-FREQUENCY OUTPUT. Applies power to the instrument when rotated from the OFF position and controls line-frequency voltage supplied to LINE-FREQ terminals.

6. OUTPUT and GND. Terminals for connection to the selected frequency output.

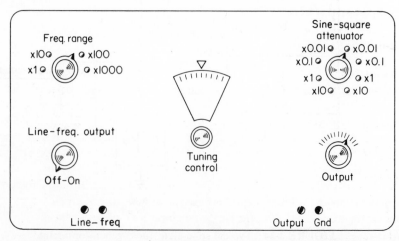

FIGURE 2-35 Typical audio-generator controls and indicators

7. LINE-FREQ. Terminals provide a line-frequency voltage adjustable up to 6 V.

Typical audio-generator operating procedures. The following steps describe the basic operating procedures for a typical audio generator:

1. Observe the general safety precautions described in Sec. 2-1.

2. Turn the OFF-ON/LINE-FREQ OUTPUT switch to OFF, and connect the power cord to an ac outlet.

3. Turn the OFF-ON/LINE-FREQ OUTPUT switch to the ON position, and allow 15 minutes (or whatever time is recommended by the manufacturer) for warm-up.

4. Connect a set of test leads to the OUTPUT and GND terminals.

5. Adjust the TUNING dial to the basic setting for the desired frequency.

6. Set the FREQ RANGE switch to the appropriate multiple. For example, for a tuning range of 20 to 200 Hz, set the FREQ RANGE switch to × 1; for a tuning range of 200 to 2,000 Hz, set the FREQ RANGE switch to × 10; and so on.

7. Set the SINE-SQUARE function switch to obtain the desired wave form and amplitude. The four left-hand settings are for sine-wave, and the four right-hand settings are for square wave. The OUTPUT control can be used for further regulation of the output voltage.

8. Clip the cable attached to the GND terminal of the audio generator to the ground of the circuit being serviced. Connect the probe or clip of the OUTPUT terminal to the point of signal injection.

2-6. OSCILLOSCOPES

The cathode-ray oscilloscope (CRO) is an extremely fast *x-y* plotter capable of plotting an input signal versus another signal or versus time, whichever is required. A luminous spot acts as a stylus or pen and moves over the display area in response to input voltages. The formal name *cathode-ray oscilloscope* is usually shortened to *oscilloscope* (or simply *scope*). An *oscillosgraph* is the pictorial representation of an oscilloscope trace. However, some older texts apply the word *oscillograph* to the complete piece of equipment.

In most oscilloscope applications, the *y* axis (vertical) input receives its signal from the voltage being examined, moving the luminous spot up or down in accordance with the instantaneous value of the voltage. The *x* axis (horizontal) input is usually an internally generated linear ramp voltage that moves the spot uniformly from left to right across the display screen. The spot then traces a *curve* that shows how the input voltage varies as a function of time.

If the signal being examined is repetitive at a fast-enough rate, the display on an oscilloscope screen appears to stand still. The oscilloscope is thus a means of visualizing *time-varying voltages* or *signals*. As such, the oscilloscope has become a universal troubleshooting tool for all types of electronic equipment.

In addition to the basic oscilloscope, there are many highly-specialized types of scopes. For example, the *storage oscilloscope* is used to display and hold *one-time* wave forms, and the *sampling oscilloscope* is used to display *extremely fast* wave forms. Because of the special circuits used in these instruments, and because there is no standardization in operating controls and indicators, we will make no attempt to discuss scope circuits and controls here. Instead, we will concentrate on using oscilloscopes as troubleshooting tools. That is, we will show how typical scope controls and indicators are used to make the measurements required during troubleshooting of basic circuits.

2-6.1 Oscilloscope Operating Precautions

In addition to the general safety precautions described in Sec. 2-1, the following specific precautions should be observed when operating any type of oscilloscope:

1. Even if you have had considerable experience with scopes, always study the instruction manual of any oscilloscope with which you are not familiar.

2. Use the procedures outlined in Sec. 2-6.2 to place the oscilloscope in operation. It is good practice to go through the procedures each time that the oscilloscope is used. This is especially true when the scope is used by other persons. The operator cannot be certain that position, focus, and especially, intensity controls are at safe positions, and the oscilloscope's cathode-ray tube (CRT) could be damaged by switching it on immediately.

3. As in the case of any CRT device (such as a television receiver), the CRT spot should be *kept moving* on the screen. If the spot must remain in one position, keep the intensity control as low as possible.

4. Always use the minimum intensity necessary for good viewing.

5. If at all possible, avoid using an oscilloscope in direct sunlight or in a brightly lighted room. This will permit a low-intensity setting. When the scope must be used in bright light, use the viewing hood.

6. Make all measurements in the center area of the screen. Even if the CRT is flat, there is a chance of reading errors caused by distortion at the edges.

7. Use only shielded probes. Never allow your fingers to slip down to the metal probe tip when the probe is in contact with a hot circuit.

8. Avoid operating an oscilloscope in strong magnetic fields. Such fields can cause distortion of the display. Most quality oscilloscopes are well shielded against magnetic interference. However, the face of the CRT is still exposed and is subject to magnetic interference.

9. Most oscilloscopes and their probes have some maximum input voltage specified in the instruction manual. Do not exceed this maximum. Also, do not exceed the maximum line voltage or use a different power frequency.

10. Avoid operating the scope with the shield or case removed. Besides the danger of exposing high-voltage circuits (several thousand volts are used in the CRT), there is the hazard that the CRT will implode and scatter glass at high velocity.

11. Avoid vibration and mechanical shock. Like most electronic equipment, an oscilloscope is a delicate instrument.

12. If an internal fan or blower is used in the scope, make sure that it is operating. Keep ventilation air filters clean.

13. Do not attempt repair of a scope unless you are a qualified instrument technician. If you must adjust any internal circuits, follow the instruction manual.

14. Study the circuit that is to be tested before making any test connections. Try to match the capabilities of the scope to this circuit. For example, if the circuit has a range of measurements to be made (alternating current, direct current, radio frequency, pulse), you must use a wideband dc oscilloscope with a low-capacitance probe (sec. 2-7) and possibly a demodulator probe. Do not try to measure 3 MHz signals with a 100 kHz bandwidth scope. On the other hand, it is wasteful to use a 50 MHz dual-trace laboratory scope to check out the audio sections of a transistor radio.

15. The most important oscilloscope operating precautions are summarized in Fig. 2-36.

2-6.2 Placing an Oscilloscope in Operation

After you have digested the oscilloscope manual's instructions for setting up the equipment, compare them with the following *general* or *typical* procedures. Always follow the manual's procedures in the case of conflicting instructions.

1. Set the power switch to OFF.

2. Set the internal recurrent sweep to OFF.

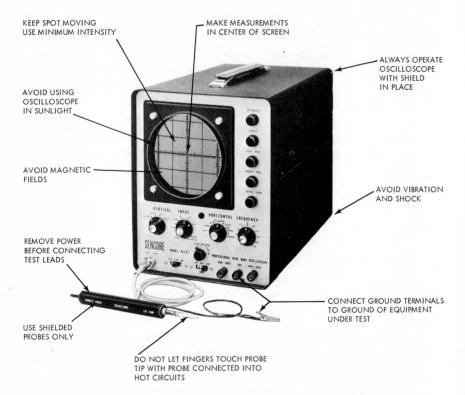

KEEP SPOT MOVING
USE MINIMUM INTENSITY

MAKE MEASUREMENTS
IN CENTER OF SCREEN

ALWAYS OPERATE
OSCILLOSCOPE
WITH SHIELD
IN PLACE

AVOID USING
OSCILLOSCOPE
IN SUNLIGHT

AVOID MAGNETIC
FIELDS

AVOID VIBRATION
AND SHOCK

REMOVE POWER
BEFORE CONNECTING
TEST LEADS

CONNECT GROUND TERMINALS
TO GROUND OF EQUIPMENT
UNDER TEST

USE SHIELDED
PROBES ONLY

DO NOT LET FINGERS TOUCH PROBE
TIP WITH PROBE CONNECTED INTO
HOT CIRCUITS

FIGURE 2-36 Summary of oscilloscope operating precautions (Sencore)

3. Set the FOCUS, GAIN, INTENSITY, and SYNC controls to their lowest position (usually a full turn counterclockwise).

4. Set the SWEEP selector to EXTERNAL.

5. Set the VERTICAL and HORIZONTAL POSITION controls to their approximate midpoint.

6. Set the POWER switch to ON. It is assumed that the power cord has been connected. This is always a good idea.

7. After a suitable warm-up period (as recommended by the manual), adjust the INTENSITY control until the trace spot appears on the screen. If a spot is not visible at any setting of the INTENSITY control, the spot is probably off screen (unless the scope is defective). If necessary, use the VERTICAL and HORIZONTAL POSITION controls to bring the spot into view. Always use the lowest setting of the INTENSITY control needed to see the spot. This will prevent burning the scope screen.

8. Set the FOCUS control for a sharp, fine dot.

9. Set the VERTICAL and HORIZONTAL POSITION controls to center the spot on screen.

10. Set the SWEEP selector to INTERNAL. If more than one internal sweep is available, this should be the LINEAR INTERNAL sweep.

11. Set the internal recurrent sweep to ON. Set the SWEEP frequency to any frequency or recurrent rate higher than 100 Hz.

12. Adjust the HORIZONTAL GAIN control, and check that the spot is expanded into a horizontal trace or line. The line length should be controllable by adjusting the HORIZONTAL GAIN control.

13. Return the HORIZONTAL GAIN control to zero (or its lowest setting). Set the internal recurrent sweep to OFF.

14. Set the VERTICAL GAIN control to the approximate midpoint. Touch the VERTICAL input with your finger. The stray signal pickup should cause the spot to be deflected vertically into a trace or line. Check that the line length is controllable by adjustment of the VERTICAL GAIN control.

15. Return the VERTICAL GAIN control to zero (or its lowest setting).

16. Set the internal recurrent sweep to ON. Advance the HORIZONTAL GAIN control to expand the spot into a horizontal line.

17. If required, connect a probe to the VERTICAL input.

18. The oscilloscope should now be ready for immediate use. Depending on the test to be performed, the scope may require *calibration*. Typical voltage-calibration procedures are described in later sections of this chapter.

2-6.3 *Measuring Wave-Form Voltage with an Oscilloscope*

The oscilloscope has both advantages and disadvantages when used to measure the voltage of wave forms or signals in electronic circuits during troubleshooting. The most obvious advantage is that the scope shows wave form, frequency and/or time duration, and phase simultaneously with the amplitude of the voltage being measured. The meters described in Secs. 2-2 and 2-3 show only amplitude. If the only value required when troubleshooting a particular circuit is voltage (or current), use the meter because of its simplicity in readout. However, when wave-shape characteristics are of equal importance to amplitude, use the scope.

This section describes *typical* procedures for voltage (and current) measurements and *typical* calibration procedures. Keep in mind that you must relate these procedures to the specific oscilloscope you use for troubleshooting.

Peak-to-peak measurements with ac laboratory scopes

1. Connect the equipment, as shown in Fig. 2-37.

2. Place the scope in operation (Sec. 2-6.2).

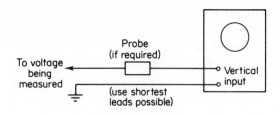

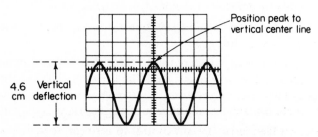

FIGURE 2-37   Measuring peak-to-peak voltages

3. Set the VERTICAL step attenuator to a deflection factor that will allow the *expected* signal amplitude to be displayed without overdriving the vertical amplifier.

4. Set the INPUT selector to measure alternating cur.ent. Connect the probe to the signal being measured.

5. Switch on the oscilloscope's internal recurrent sweep.

6. Adjust the sweep FREQUENCY for several cycles on the screen.

7. Adjust the HORIZONTAL GAIN control to spread the pattern over as much of the screen as desired.

8. Adjust the VERTICAL POSITION control so that the downward excursion of the wave form coincides with one of the screen lines below the center line, as shown in Fig. 2-37.

9. Adjust the HORIZONTAL POSITION control so that one of the upper peaks of the signal lies near the vertical center line.

10. Measure the peak-to-peak vertical deflection in divisions (which are usually in centimeters (cm) on a laboratory scope).

11. Multiply the distance measured in step 10 by the VERTICAL step attenuator setting. Also include the attenuation factor of the probe (if any).

For example, assume a peak-to-peak vertical deflection of 4.6 divisions (Fig. 2-37), using a ×10 attenuator probe, and a vertical deflection factor of 0.5 V per division:

Volts (peak-to-peak) = 4.6 × 0.5 × 10 = 23 V

If the wave form being measured is a sine wave, the peak-to-peak value can be converted to peak, root-mean-square, or average, as shown in Fig. 2-11, and vice versa.

Peak-to-peak measurements with commercial ac scopes. The procedures for measuring wave-form voltage with commerical scopes are essentially the same as those used for laboratory scopes, with one possible exception. Commercial scopes usually have a VERTICAL GAIN control; whereas laboratory scopes usually have a VERTICAL step attenuator. The VERTICAL GAIN control must be set to a given CALIBRATE position, as determined during the calibration procedure. Such procedures are described in the following paragraphs.

Calibrating the vertical amplifier for voltage measurements. On those laboratory scopes that have a vertical step attenuator related to some specific deflection factor (such as 5 V/cm), the calibration procedure is an internal adjustment done as part of routine maintenance. On commercial scopes, the vertical amplifier must be calibrated for voltage measurements.

The basic procedure consists of applying a reference voltage of known amplitude to the VERTICAL input and adjusting the VERTICAL GAIN control for specific deflection. Then, the reference voltage is removed, and the test voltages are measured, *without changing* the setting of the VERTICAL GAIN control. The calibration will remain accurate as long as the VERTICAL GAIN control is at this CALIBRATE position.

The reference voltage may be external or internal and may be direct current, alternating current, or square wave. On scopes that do not have an internal voltage-reference source, it is necessary to use an external signal of *known accuracy.* Most modern scopes have an internal voltage source (sine wave or square wave) of known amplitude and accuracy available for calibration. On some scopes, this calibrating voltage is available from terminals or a jack on the front panel. On other scopes, the calibrating voltage is applied to the vertical input when one of the controls (usually the VERTICAL INPUT selector) is set to CALIBRATE or the CAL position.

The following steps describe the calibration of vertical amplifier voltage for a typical commercial scope. It is assumed that the reference voltage is alternating current and that the reference-voltage value is such that it will produce *near full-scale deflection* with the VERTICAL GAIN control near mid-scale.

The accuracy of the scope's voltage measurements will be no greater than the accuracy of the reference voltage. Also, the scope display is usually calibrated for peak-to-peak voltage; whereas the meter or other device indicating the reference voltage will probably be in the rms value. If the reference voltage is a sine wave, the rms value can be converted to peak-to-peak, as shown in Fig. 2-11. If the reference voltage is a square wave or pulse, its value is peak-to-peak.

1. Connect the equipment, as shown in Fig. 2-38.

2. Place the scope in operation (sec. 2-6.2).

3. Using the VERTICAL POSITION control (not gain control), position the trace to the horizontal center line. Switch the internal recurrent sweep to ON (for a normal line trace) or to OFF (for a dot), whichever is most convenient for calibration. If the internal sweep is not on, the dot trace will appear as a vertical line when the reference voltage is applied.

4. Apply the reference voltage, and adjust it to the desired value. The exact value of the reference voltage depends upon the scope scale. For example, if there are 10 vertical divisions (5 above and 5 below the horizontal center line), a value of 1, 10, or 100 V is convenient.

5. *Without touching* the VERTICAL POSITION control, adjust the VERTICAL GAIN control to align the tips of positive half-cycles and tips of negative half-cycles with the desired scale divisions.

   For example, assuming a reference voltage of 10 V peak-to-peak and a scale like the one shown in Fig. 2-38, set the VERTICAL GAIN control so that the trace is spread from the top line (5 divisions up from the center line) to the bottom line (5 divisions below the center line).

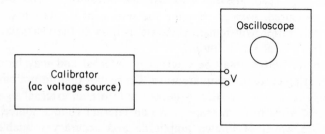

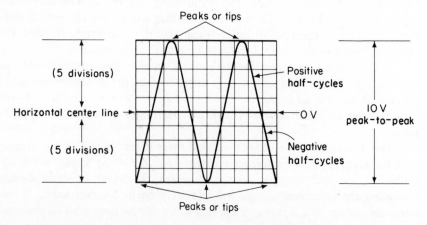

FIGURE 2-38   Voltage calibration with external alternating current

Thus, each division equals 1 V peak-to-peak. In the example shown in Fig. 2-38, this gives the scope a vertical deflection factor of 1 V per division.

If the external source is fixed, then the process must be reversed, and a *scale factor* must be selected to match the voltage. For example, assume that the only reference source is a 1 V signal (1 V rms) of known accuracy. This 1 V rms signal is equal to 2.828 V peak-to-peak, as shown in Fig. 2-11. The VERTICAL GAIN control can then be set to provide a spread (positive peak to negative peak) of slightly less than 3 divisions (2.828 divisions). This will still give a vertical deflection factor of 1 V per division.

6. If the reference-voltage source is variable, check the accuracy of the calibration by applying various voltages.

7. Note the VERTICAL GAIN control setting, and record that setting as the CALIBRATE position. Use the same position for all future voltage measurements. It is recommended that the calibration be checked at frequent intervals. Once the CALIBRATE position has been established, the trace can be moved up or down (as required) by the VERTICAL POSITION control without affecting the volts-per-division factor.

Composite and pulsating voltage measurements. In most troubleshooting applications, the voltages measured are *composites* of alternating current and direct current or are pulsating direct current. For example, a transistor amplifier used to amplify an ac signal has both alternating current (the signal) and direct current (the power source) on its collector. Such composite and pulsating voltages can be measured quite easily on a scope capbable of measuring direct current (an ac scope will show only the ac portion of the composite voltage).

The procedures for measuring a composite voltage are essentially a combination of peak-to-peak measurements and instantaneous dc measurements, as discussed in the following steps:

1. Connect the equipment, as shown in Fig. 2-39.

2. Place the scope in operation (Sec. 2-6.2).

3. Set the VERTICAL step attenuator to a deflection factor that will allow the expected signal, *plus any direct current*, to be displayed without overdriving the vertical amplifier.

4. Set the INPUT selector to GROUND.

5. Switch on the internal recurrent sweep. Adjust the HORIZONTAL GAIN control to spread the trace over as much of the screen as desired.

6. Using the VERTICAL POSITION control, move the trace to a convenient location on the screen. If the voltage to be measured is a composite and

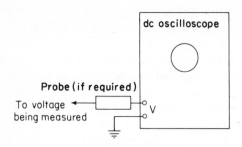

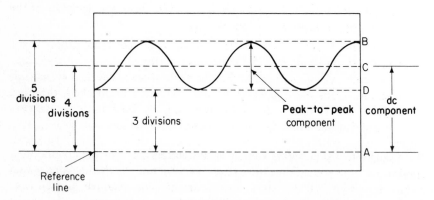

FIGURE 2-39 Measurement of composite voltages

the *average signal* (alternating current plus direct current) is positive, move the trace below the center line, as shown in Fig. 2-39. If the average is negative, move the trace above the center line. *Do not* move the VERTICAL POSITION control after this reference has been established.

7. Set the INPUT selector to measure direct current. Connect the probe to the signal being measured.

8. If the wave form is now outside the viewing area, set the VERTICAL step attenuator so that the wave form is visible.

9. Adjust the SWEEP FREQUENCY and HORIZONTAL GAIN controls to display the desired wave form.

10. Establish the polarity of the signal. Any signal-inverting switches on the scope must be in the normal position. If the wave form is above the reference line, the voltage is positive; if it is below the line, the voltage is negative.

11. Measure the distance in divisions between the reference line and the point on the wave form at which the level is to be measured. With

composite of direct current and alternating current, the trace may be on either side of the reference line or it may cross over the reference, but it usually remains on one side, as shown in Fig. 2-39.

12. Multiply the distance measured in step 11 by the VERTICAL step attenuator setting. Also include the attenuation factor of the probe (if any).

For example, assume that the vertical distance measured is 3 divisions from the reference line to point D, 4 divisions to point C, and 5 divisions to point B. Also assume that the attenuator probe is $\times 10$ and that there is a vertical deflection factor of 2 V per division for the scope.

Because the wave form is above the reference line, the dc component must be positive. The dc component (reference line to point C) is $4 \times 2 \times 10 = +80$ V.

The peak-to-peak value of the ac component (point B to D) is $2 \times 2 \times 10 = 40$ V (peak-to-peak).

Thus, there is a 40 V signal riding on a $+80$ V dc line.

2-6.4 Measuring Time and Frequency with an Oscilloscope

An oscilloscope is the ideal tool for measuring time and frequency of voltages during troubleshooting. If the horizontal sweep is calibrated directly in relation to time, such as 5 milliseconds (mS) per division, the time duration of voltage wave forms may be measured directly on the scope screen without calculation. If the time duration of *one complete cycle* is measured, frequency can be calculated by simple division because frequency is the reciprocal of time duration of one cycle.

The time-frequency measurements differ somewhat for commercial and laboratory scopes. For that reason, we shall describe all time-frequency measurements likely to be found in basic troubleshooting.

Time-duration measurements with laboratory oscilloscopes. The horizontal-sweep circuit of a laboratory scope is usually provided with a selector control that is direct-reading in relation to time; that is, each horizontal division on the scope screen has a definite relation to time at a given position of the horizontal sweep rate switch (such as milliseconds per centimeter or microseconds per centimeter). With such scopes, the wave form can be displayed, and the time duration of the complete wave form (or any portion) can be measured directly, as follows:

1. Connect the equipment, as shown in Fig. 2-40.

2. Place the scope in operation (Sec. 2-6.2).

3. Set the VERTICAL step attenuator to a deflection factor that will allow the expected signal to be displayed without overdriving the vertical amplifier.

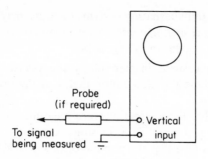

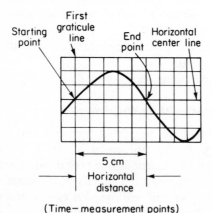

FIGURE 2-40 Measuring time duration

4. Connect the probe (if any) to the signal being measured.

5. Switch on the internal recurrent sweep. Set the HORIZONTAL SWEEP control to the *fastest* sweep rate that will display a *convenient* number of divisions between the time-measurement points, as shown in Fig. 2-40. On most scopes, it is recommended that the extreme sides of the screen not be used for time-duration measurements. There may be some nonlinearity at the beginning and end of the sweep.

6. Adjust the VERTICAL POSITION control to move the points between which the time measurement is made to the horizontal center line.

7. Adjust the HORIZONTAL POSITION control to move the *starting point* of the time-measurement area to the first (vertical) screen line. If the HORIZONTAL SWEEP is provided with a variable control, make certain it is set on OFF (or in the CALIBRATE position).

8. Measure the horizontal distance between the time-measurement points.

9. Multiply the distance measured in step 8 by the setting of the HORIZONTAL SWEEP control.

For example, assume that the distance between the time-measurement points is 5 divisions (Fig. 2-40) and that the HORIZONTAL SWEEP control is set to 0.1 mS per division. The time duration is $5 \times 0.1 =$ 0.5 mS, or 0.0005 second.

Frequency measurement with a laboratory scope. The frequency measurement of a periodically recurrent wave form is essentially the same as time-duration measurement, except that an additional calculation must be performed. In effect, a time-duration measurement is made, then the time duration is divided into 1 because frequency is the reciprocal of one cycle.

Figure 2-41 shows the scope measurement for complete and incomplete cycles of typical wave forms found in troubleshooting. Figures 2-41(a) and

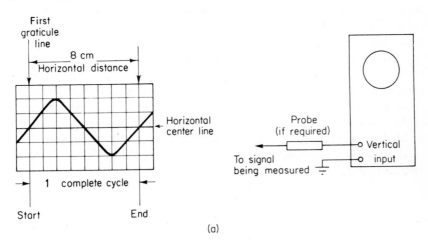

(a)

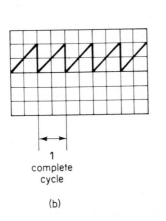

(b)

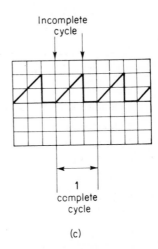

(c)

FIGURE 2-41 Measuring frequency where horizontal sweep is calibrated in relation to time

2-41(b) show complete cycles of an approximate sine wave and a sawtooth wave, respectively. To find the frequency of either signal, divide the measured time into 1. For example, assume that the HORIZONTAL SWEEP control is set to 0.1 mS per division in each case.

The distance between the beginning and end of a complete cycle for the wave form of Fig. 2-41(a) is 8 divisions. Thus, one complete cycle is 0.8 mS (or 0.0008 second), and the frequency is 1/0.0008 = 1,250 Hz.

The distance between the beginning and end of a complete cycle for the wave form of Fig. 2-41(b) is 2 divisions, one complete cycle is 0.2 mS, and the frequency is 1/0.0002 = 5,000 Hz, or 5 kHz.

In the case of the wave form in Fig. 2-41(c), the complete cycle is 3 divisions (0.3 mS) and must not be confused with the incomplete or partial cycle shown as 2 divisions. With a complete cycle at 3 divisions (0.3 mS time duration), the frequency is 1/0.0003 = 3,333 Hz.

Frequency and time measurements with commercial oscilloscopes. The horizontal-sweep circuit of most commercial oscilloscopes is provided with controls that are direct-reading in relation to frequency. Usually, there are two controls: a STEP selector and a VERNIER or VARIABLE. The sweep frequency of the horizontal trace is equal to the scale settings of the controls. Thus, when a signal is applied to the vertical input and the horizontal sweep controls are adjusted until one complete cycle occupies the entire length of the trace, the vertical signal is equal in frequency to the scale settings of the HORIZONTAL SWEEP control. If desired, the frequency can then be converted to time:

$$\text{Time} = \frac{1}{\text{frequency}}$$

1. Connect the equipment, as shown in Fig. 2-42.

2. Place the oscilloscope in operation (Sec. 2-6.2).

3. Set the VERTICAL step attenuator to a deflection factor that will allow the expected signal to be displayed without overdriving the vertical amplifier.

4. Connect the probe (if any) to the signal being measured.

5. Switch on the internal recurrent sweep. Set the HORIZONTAL SWEEP controls (STEP and VERNIER) so that one complete cycle occupies the *entire* length of the trace.

6. Read the unknown vertical signal frequency directly from the HORIZONTAL SWEEP FREQUENCY control settings.

For example, assume that the STEP HORIZONTAL SWEEP control is set to the 10 kHz position and that the VERNIER SWEEP control indicates 5 (on a total scale of 10). This indicates that the horizontal-sweep frequency is 5 kHz. If one complete cycle of vertical signal occupies the

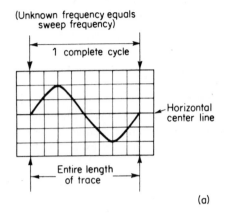

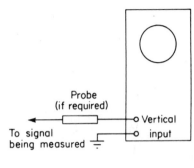

(a)

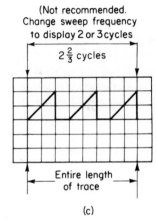

(b)   (c)

FIGURE 2-42   Measuring frequency where horizontal sweep is calibrated directly in units of frequency

entire length of the trace, as shown in Fig. 2-42(a), the vertical signal is also at a frequency of 5 kHz.

7. If it is not practical to display one cycle on the trace, more than one cycle can be displayed, and the resultant horizontal-sweep frequency indication multiplied by the number of cycles.

You must remember two important points. First, multiply the indicated sweep frequency by the number of cycles appearing on the trace.

For example, assume (once again) that the horizontal-sweep frequency is 5 kHz but that the vertical signal is *four times* that amount; (that is, four complete cycles of vertical signal occupy the entire trace, as shown in Fig. 2-42(b). Then, the signal frequency is 5 kHz × 4, or 20 kHz.

Second, it is absolutely essential that an *exact* number of cycles occupies the entire length of the trace.

For example, assume that the horizontal-sweep frequency is again 5 kHz and that the $2\frac{2}{3}$ cycles of vertical occupy the entire length of the trace, as shown in Fig. 2-42(c). This indicates a frequency of 13.3 kHz. However, the exact percentage of the incomplete cycle (one-third) is quite difficult to measure. It is far simpler and more accurate to increase or decrease the horizontal-sweep frequency until an exact number of cycles appears.

2-7. PROBES

In practical troubleshooting, all meters and oscilloscopes operate with some type of probe. In addition to providing for electrical contact to the circuit being tested, probes serve to modify the voltage being measured to a condition suitable for display on an oscilloscope or readout on a meter.

For example, assume that a very high voltage must be measured and that this voltage is beyond the maximum input limits of the meter or scope. A *voltage-divider probe* can be used to reduce the voltage to a safe level for measurement. Under these circumstances, the voltage is reduced by a fixed amount (known as the *attenuation factor*), usually on the order of 10:1 or 100:1.

2-7.1 Basic Probe

In its simplest form, the basic probe is a *test prod*. In physical appearance, the probe is a thin metal rod connected to the meter or scope input through an insulated flexible lead. The entire rod *except* for the tip is covered with an insulated handle so that the probe can be connected to any point of the circuit without touching nearby circuit parts. Sometimes, the probe tip is provided with an alligator clip so that it is not necessary to hold the probe at the circuit point.

Such probes work well on circuits carrying dc and audio signals. However, if the alternating current is at a high frequency, or if the gain of the meter (such as an electronic meter) or scope amplifier is high, it may be necessary to use a special *low-capacitance probe*. Hand capacitance in a simple probe or test prod can cause hum pickup, particularly if amplifier gain is high. This condition can be offset by shielding in low-capacitance probes. More important, however, is the fact that the input impedance of the meter or scope is connected directly to the circuit being tested when a simple probe is used.

Such impedance can disturb circuit conditions (as discussed in Secs. 2-2 and 2-3).

2-7.2 Low-Capacitance Probes

The basic circuit of a low-capacitance probe is shown in Fig. 2-43. The series resistance R_1 and capacitance C_1, as well as the parallel or shunt R_2, are surrounded by a shielded handle. The values of R_1 and C_1 are preset at the factory by screwdriver adjustment and should not be disturbed unless recalibration is required, as discussed in Sec. 2-7.8.

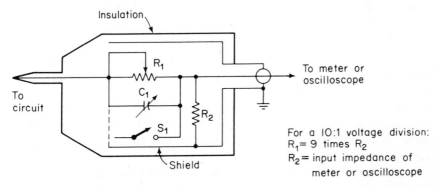

For a 10:1 voltage division:
R_1= 9 times R_2
R_2 = input impedance of meter or oscilloscope

FIGURE 2-43   Typical low-capacitance probe circuit

In most low-capacitance probes, the values of R_1 and R_2 are selected to form a 10:1 voltage divider between the circuit being tested and the meter or scope input. Thus, the probes serve the dual purpose of capacitance reduction and voltage reduction. You should remember that voltage indications will be one-tenth (or whatever value of attenuation is used) of the actual value when such probes are connected at the inputs of meters or scopes.

The capacitance value of C_1 in combination with the values of R_1 and R_2 also provide a capacitance reduction, usually in the range of 3:1 to 11:1.

There are probes that combine the features of low-capacitance probes and the basic probe (or test prod). In such probes, a switch (shown as S_1 in Fig. 2-43) is used to short both C_1 and R_1 when a direct input (simple test prod) is required. With S_1 open, both C_1 and R_1 are connected in series with the input, and the probe provides the low-capacitance and voltage-division features.

2-7.3 Resistance-Type Voltage-Divider Probes

A resistance-type voltage-divider probe is used when the primary concern is reduction of voltage. The resistance-type probe, shown in Fig. 2-44, is similar to the low-capacitance probe described in Sec. 2-7.2 except that the frequency-compensating capacitor is omitted.

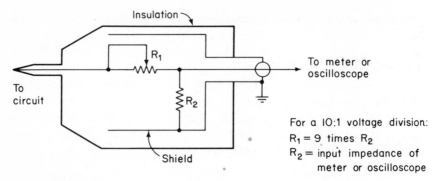

For a I0:1 voltage division:
$R_1 = 9$ times R_2
$R_2 =$ input impedance of meter or oscilloscope

FIGURE 2-44   Typical resistance-type voltage-divider probe circuit

Usually, the simple resistance-type probe is used when a voltage reduction of 100:1 or greater is required and when a flat frequency response is of no particular concern.

As shown in Fig. 2-44, the values of R_1 and R_2 are selected to provide the necessary voltage division and to match the input impedance of the meter or scope. Resistor R_1 is usually made variable so that an exact voltage division can be obtained.

Because of their voltage-reduction capabilities, resistance-type probes are often known as *high-voltage probes*. Some resistance-type probes are capable of measuring potentials on the order of 50 kilovolts (kV) with a 1,000:1 voltage reduction.

2-7.4 Capacitance-Type Voltage-Divider Probes

In certain isolated cases, the resistance-type voltage-divider probes described in Sec. 2-7.3 are not suitable for measurement of high voltages because stray conduction paths are set up by the resistors. A capacitance-type probe, shown in Fig. 2-45, can be used in those cases.

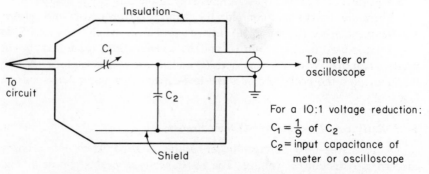

For a I0:1 voltage reduction:
$C_1 = \frac{1}{9}$ of C_2
$C_2 =$ input capacitance of meter or oscilloscope

FIGURE 2-45   Typical capacitance-type voltage-divider probe circuit '

In the capacitance probe shown in Fig. 2-45, the values of C_1 and C_2 are selected to provide the necessary voltage division and to match the input capacitance of the meter or scope. Capacitor C_1 is usually made variable so that an exact voltage division can be obtained.

2-7.5 RF probes

When the signals to be measured are at radio frequencies and are beyond the frequency capabilities of the meter or scope circuits, an RF probe is required. RF probes convert (rectify) the RF signals to a dc output voltage that is equal to the peak RF voltage. The dc ouput of the probe is then applied to the meter or scope input and is displayed as a voltage readout in the normal manner. In some RF probes, the dc output is equivalent to peak RF voltage; whereas in other probes, the readout is equal to rms voltage.

The basic circuit of an RF probe is shown in Fig. 2-46. This circuit can be used to provide either peak output or rms output. Capacitor C_1 is a high-capacitance dc blocking capacitor used to protect diode CR_1. Usually, a germanium diode is used for CR_1, which rectifies the RF voltage and produces a dc output across R_1. In some probes, R_1 is omitted so that the dc voltage is developed directly across the input circuit of the meter or scope. This dc voltage is equal to the peak RF voltage, less whatever forward drop exists across the diode CR_1.

When a dc output voltage equal to the root-mean-square of the RF voltage is wanted, a series-dropping resistor (shown as R_2 in Fig. 2-46) is added to the circuit. Resistor R_2 drops the dc output voltage to a level that equals 0.707 of the peak RF value.

The RF probe shown in Fig. 2-46 is a half-wave probe. A full-wave probe provides an output to the meter or scope that is (approximately) equal to the peak-to-peak value of the voltage being measured. This is particularly

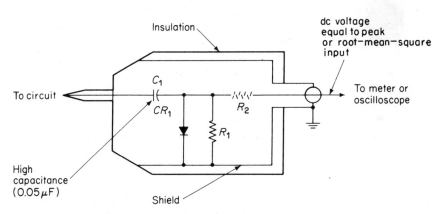

FIGURE 2-46 Typical half-wave RF probe circuit.

important when measuring pulses, square waves, and any other complex wave form.

A full-wave probe circuit is shown in Fig. 2-47. Full-wave probes are usually found with meters rather than scopes. Because most meters are calibrated to read in rms values, the probe output is reduced to 0.3535 of the peak-to-peak value. This is done by selecting or adjusting the value of resistor R_1. A few VTVMs and electronic meters are provided with peak-to-peak scales or readouts. These meters are used in certain laboratory applications and in television service to measure such values as horizontal-oscillator or deflection-coil voltages, input to video amplifier, and output of the vertical amplifier. On such meters, resistor R_1 is omitted so that the dc output is equivalent to peak-to-peak input.

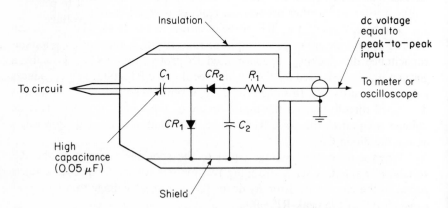

FIGURE 2-47   Typical full-wave RF probe circuit

2-7.6 Demodulator Probes

The circuit of a demodulator probe is essentially the same as that of the RF probe described in Sec. 2-7.5. But the circuit values and the basic functions are somewhat different.

Both the half-wave and full-wave probes are used to convert high-frequency signals (usually an RF carrier) to a dc voltage that can be measured on a meter or scope. When the high-frequency signals contain ac or pulsating dc modulation (such as a modulated RF carrier), a demodulator probe is more effective for signal tracing.

The basic circuit of the demodulator probe is shown in Fig. 2-48. Here, capacitor C_1 is a *low-capacitance* dc blocking capacitor. (In the RF probe, a high-capacitance value is required for C_1 to ensure that the diode operates at the peak of the RF signal. This is not required for a demodulator probe.) Germanium diode CR_1 demodulates (or detects) the AM signal and produces voltages across load resistor R_1. (C_1 and R_1 also act as a filter.)

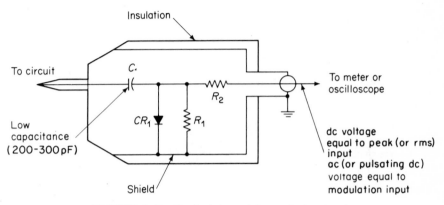

FIGURE 2-48   Typical demodulator probe circuit

The demodulator probe produces ac and dc outputs. The RF carrier frequency is converted to a dc voltage equal to the peak of the RF carrier. The low-frequency modulating voltage appears as alternating current (or pulsating direct current) at the probe output.

In troubleshooting applications, the meter or scope is set to measure direct current, and the RF carrier is measured. Then the meter or scope is set to measure alternating current, and the modulating voltage is measured. Resistor R_2 is used primarily for isolation between the circuit being tested and the meter or scope input. Resistor R_2 can also serve as a calibrating resistance so that the output will be of equal value (root-mean-square, peak, and so on). However, in general, demodulator probes are used primarily for *signal tracing* (as part of troubleshooting), and their output is not calibrated to any particular value.

2-7.7 Special-Purpose Probes

The probes described thus far are in common use as troubleshooting tools. There are also a number of special probes. The transistorized signal-tracing probe and the cathode-follower probe are typical examples.

Cathode-follower probe. The cathode-follower probe provides a means of connecting into a circuit without disturbing circuit operation. A cathode-follower probe provides a high-impedance input (as does an electronic meter, Sec. 2-3) that does not disturb circuit conditions and a low-impedance output to match the input of an amplifier (e.g., some amplifiers used in scopes).

The basic circuit of a cathode-follower probe is shown in Fig. 2-49. The grid-input resistor R_1 is of high resistance (usually several megohms); the cathode-load resistor R_2 is of the same value as the amplifier-input impedance. Thus, the circuit being tested has a high impedance that does not disturb the

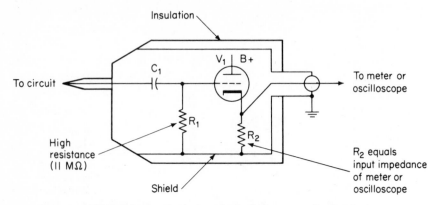

FIGURE 2-49   Typical cathode-follower probe circuit

circuit; whereas the amplifier (such as the vertical input amplifier of a scope) has a matched impedance. The one disadvantage of the cathode-follower probe is that external power is required for the tube V_1.

Transistorized signal-tracing probes. It is possible to increase the sensitivity of a probe with a transistor amplifier. Such an arrangement is particularly useful with a VOM (Sec. 2-2) for measuring small signal voltages during troubleshooting. An amplifier is usually not required for an electronic meter or scope because such instruments contain built-in amplifiers.

A transistor probe and amplifier circuit is shown in Fig. 2-50. This circuit will increase the sensitivity of the probe by at least 10:1 and should provide good response up to about 500 MHz. The circuit is not normally calibrated to provide a specific voltage indication; rather, it is used to increase the sensitivity of the probe for signal-tracing purposes.

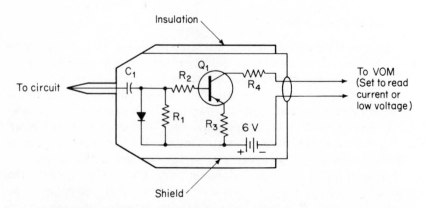

FIGURE 2-50   Typical transistorized signal-tracing probe circuit

2-7.8 Probe Compensation and Calibration

Probes must be calibrated to provide a proper output to the meter or scope with which they are to be used. Probe compensation and calibration are done at the factory and must be accomplished with proper test equipment. The following paragraphs describe the *general* procedures for compensating and calibrating probes. *Never* attempt to adjust a probe unless you follow the instruction manual and have the proper test equipment. An improperly adjusted probe will provide erroneous readings and may cause undesired circuit loading.

Probe compensation. The capacitors that compensate for excessive attenuation of high-frequency signal components (through the probe's resistance dividers) affect the entire frequency range from some mid-band point upward. (Capacitor C_1 in Fig. 2-43 is an example of such a compensating capacitor.)

Compensating capacitors must be adjusted so that the higher-frequency components are attenuated by the same amount as low frequency and direct current. It is possible to check the adjustment of the probe-compensating capacitors using a square-wave signal source. This is done by applying the square-wave signals directly to a scope input and then applying the same signals through the probe and noting any *change* in pattern. In a properly compensated probe, there should be no change (except for a possible reduction of the amplitude).

Figure 2-51 shows typical square-wave displays with the probe properly compensated, undercompensated (high frequencies underemphasized), and overcompensated (high frequencies overemphasized). Proper compensation of probes is often neglected, especially when probes are used interchangeably

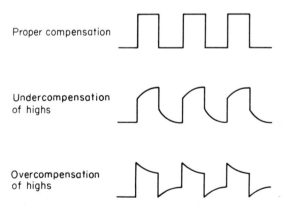

Proper compensation

Undercompensation of highs

Overcompensation of highs

FIGURE 2-51 Typical square-wave displays showing frequency compensation of probes

with meters or scopes having different input characteristics. It is recommended that any probe be checked with square-wave signals before it is used in troubleshooting.

Another problem related to probe compensation is that the *input capacitance* of the meter or scope can change with age. Also, in the case of vacuum-tube meters and scopes, the input capacitance can change when tubes are changed. Either way, all the compensated dividers can be improperly adjusted. Readjustment of the probe will not correct for the change needed by the input circuits of the meter or scope.

Probe calibration. The main purpose of probe calibration is to provide a specific output for a given input. For example, the value of resistor R_1 in Fig. 2-44 is adjusted (or selected) to provide a specific amount of voltage division. During calibration, a voltage of known value and accuracy is applied to the input. The output is monitored, and R_1 is adjusted to provide a given value. For example, on a 10:1 divider probe, 10 V can be applied to the input, and R_1 is adjusted for a 1 V output.

Or, for example, the value of resistor R_2 in Fig. 2-46 is adjusted (or selected) to provide a dc output that equals 0.707 of the peak RF input value, that is, with a 10 V peak RF input R_2 is adjusted to provide a 7.07 V dc output.

2-7.9 Probe Troubleshooting Techniques

Although a probe is a simple instrument and does not require specific operating procedures, several points should be considered to use a probe effectively in troubleshooting.

Circuit loading. When a probe is used, the *probe's* impedance (rather than the meter's or scope's impedance) determines the amount of circuit loading. As we have noted, connecting a meter or scope to a circuit may alter the amplitude or wave form at the point of connection. To prevent this, the impedance of the measuring device must be large in relation that of the circuit being tested. Thus, a high-impedance probe will offer less circuit loading, even though the meter or scope may have a lower impedance.

Measurement error. The ratio of the two impedances (of the probe and the circuit being tested) represents the amount of probable error. For example, a ratio of 100:1 (perhaps a 100 MΩ probe to measure the voltage across a 1 MΩ circuit) will account for an error of about 1 percent. A ratio of 10:1 will produce an error of about 9 percent.

Effects of frequency. The input impedance of a probe is not the same at all frequencies; it continues to get smaller at higher frequencies. (Capacitive reactance and impedance decrease with an increase in frequency.) All probes have some input capacitance. Even an increase at audio frequencies may produce a significant change in impedance.

Shielding capacitance. When using a shielded cable with a probe to minimize pickup of stray signals and hum, the additional capacitance of the cable should be recognized. The capacitance effects of a shielded cable can be minimized by terminating the cable at one end in its characteristic impedance. Unfortunately, this is not always possible with the input circuits of most meters and scopes.

Relationship of loading to attenuation factor. The reduction of loading (either resistive or capacitive) due to use of probes may not be the same as the attenuation factor of the probe. (Capacitive loading is almost never reduced by the same amount as the attenuation factor because of the additional capacitance of the probe cable). For example, a typical 5:1 attenuator probe may be able to reduce capacitive loading by about 2:1. A 50:1 attenuator probe may reduce capacitive loading by about 10:1. Beyond this point, little improvement can be expected because of stray capacitance at the probe tip.

Checking effects of the probe. When troubleshooting, it is possible to check the effect of a probe on a circuit by making the following simple test: Attach and detach another connection of similar kind (such as another probe) and observe any difference in meter reading or scope display. If there is little or no change when the additional probe is touched to the circuit, it is safe to assume that the probe has little effect on the circuit.

Probe length and connections. Long probes should be restricted to the measurement of relatively slow-changing signals (direct current and low-frequency alternating current). The same is true for long ground leads. The ground lead should be connected where no hum or high-frequency signal components exist in the ground path between that point and the signal–pick-off point.

Measuring high voltages. Avoid applying more than the rated voltage to a probe.

3

Troubleshooting Basic
Circuits and Equipment

In this chapter, we shall discuss basic circuit (and equipment) troubleshooting. That is, we describe how the basic troubleshooting techniques discussed in Chapter 1 are combined with the practical use of test equipment discussed in Chapter 2 to locate specific faults in various types of electronic circuits.

Because alignment and adjustment, as well as testing, are part of troubleshooting, we shall describe basic adjustment procedures for the circuits and equipment discussed. Where practical, the same format is used throughout this chapter. The test, alignment, and general troubleshooting approach are given first for each circuit or type of equipment. This is followed by specific examples of troubleshooting for the circuit or equipment.

In the specific troubleshooting examples, you will note that we describe both the steps that *could* be taken and the steps that *should* be taken after each measurement or observation. Sometimes, the could-be steps are almost as logical as the should-be steps. This is common in troubleshooting modern electronic equipment. A careful study of the *difference* between the could and should steps generally makes the difference between someone who knows electronics and someone who is a really good troubleshooter.

3-1. TROUBLESHOOTING NOTES

The following notes summarize practical suggestions for troubleshooting all types of electronic equipment.

3-1.1 Solid-State Servicing Techniques

Although the following techniques apply primarily to solid-state equipment, they are generally valid for vacuum-tube equipment.

148

Transient voltages. Be sure that power to the equipment is turned off or that the line cord is removed when making in-circuit tests or repairs. Transistors (and possibly diodes) can be damaged from the transient voltages developed when changing components or inserting new transistors (in addition to the possibility of shock or short circuit). In some equipment (e.g., the instant-on television sets), certain circuits may be *live* even with the power switch set to OFF. To be on the safe side, pull the power plug.

Disconnected parts. When working on solid-state equipment, do not operate the equipment with any parts, such as loudspeakers or picture-tube yokes, disconnected. If the load is removed from some transistor circuits, heavy current will be drawn, resulting in possible damage to the transistor or another part such as an audio transformer.

Sparks and voltage arcs. Avoid sparks and arcs when troubleshooting any type of equipment, particularly solid-state units. The transient voltages developed can damage some small-signal transistors. For example, when servicing a solid-state television set, use a meter and a high-voltage probe to measure the second-anode potential of the picture tube. Do not arc the second-anode lead to the chassis for a spark test, as is often done in troubleshooting vacuum-tube television sets. Such an arc in a solid-state television set can destroy the high-voltage rectifier and possibly damage the horizontal-output transistor.

Intermittent conditions. If you run into an intermittent condition and can find no fault using routine checks, try tapping (not pounding) the components (vacuum tubes, transistors, diodes, and so forth). If this does not work for solid-state components, try rapid heating and cooling. A small portable hair dryer and a spray-type circuit cooler make good heating and cooling sources, respectively.

First, apply heat; then, cool the component. The quick change in temperature will normally cause an intermittently defective component to go bad permanently. In many cases, the component will open or short, making it easy to locate.

As an alternate procedure, measure the gain of a transistor with an in-circuit transistor tester. Then, subject the transistor to rapid changes in temperature. If the suspected transistor changes its gain *drastically*, or if there is *no* change in gain, the transistor is probably defective.

In any event, do not hold a heated soldering tool directly on or very near a transistor or diode case. This will probably destroy the transistor or diode.

If time permits, you can locate an intermittently defective transistor by measuring in-circuit gain when the equipment is cold. Then, let the equipment operate until the trouble occurs, and measure the gain of the transistors while they are hot. Some variation will be noted in all transistors, but a leaky transistor has a much lower gain reading when it is hot.

Operating control settings. If any transistor or vacuum-tube element appears to have a short (particularly the base or grid) check the settings of any operating controls or adjustment controls associated with the circuits. For example, in an audio amplifier circuit, a gain or volume control set to ZERO or MINIMUM can give the same indication as a short from element to ground. This condition is shown in Fig. 3-1.

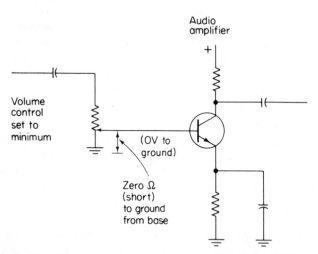

FIGURE 3-1 Example of how operating-control settings can affect voltage and resistance readings at transistor elements during troubleshooting

Record gain readings. If you must service any particular make or model of equipment regularly, record the transistor-gain readings of a unit that is working properly on the schematic for future reference. Compare these gain readings with the minimum values listed in the service literature.

Shunting capacitors. It is common practice in troubleshooting vacuum-tube circuits to shunt a suspected capacitor with a known-good capacitor. This technique is good only if the suspected capacitor is open. The test is of little value if the capacitor is leaking or shorted. In any event, avoid the shunting technique when troubleshooting solid-state circuits. This is especially true with an electrolytic capacitor (often used as an emitter bypass in solid-state circuits). The transient voltage surges can damage transistors. In general, avoid any short-circuit tests with solid-state equipment.

Test connections. Most metal-case transistors have their case tied to the collector. Thus, you can use the case as a test point. Avoid using a clip-type probe on transistors. Also avoid clipping onto some of the subminiature resistors used in solid-state equipment. Any subminiature components can break if they are handled roughly.

Injecting signals. When injecting a signal into a circuit (e.g., base of a transistor, grid of vacuum tube, input of an IC), make sure that there is a blocking capacitor in the signal-generator output. As discussed in Chapter 2, most signal generators have some form of blocking capacitor to isolate the output circuit from the dc voltages that may appear in the circuit. In the case of solid-state equipment, the blocking capacitor also prevents the base from being returned to ground (through the generator's output circuit) or from being connected to a large dc voltage (in the generator circuit). Either of these conditions can destroy the transistor. If the generator does not have a built-in blocking capacitor, connect a capacitor between the generator's output lead and the transistor base (or some other point of signal injection in the circuit).

3-1.2 Measuring Voltages in Circuit

As discussed in Chapter 1, it is possible to locate many defects in electronic circuits by measuring and analyzing voltages at the elements of active devices (e.g., grid, cathode, and plate of vacuum tubes; base, emitter, and collector of transistors). This can be done with the circuit operating and without disconnecting any parts. Once located, the defective part can be disconnected and tested or substituted, whichever is most convenient.

Vacuum-tube circuits can be analyzed with a simple VOM or electronic meter. The normal relationships of vacuum-tube elements are generally fixed. For example, the plate is positive; the cathode is at ground or positive; the grid is (usually) negative.

Transistor circuits are best analyzed with an electronic voltmeter or with a very sensitive VOM. A number of manufacturers produce VOMs designed specifically for transistor troubleshooting. (The Simpson Model 250 is a typical example.) These VOMs have very low voltage scales to measure the differences that often exist between elements of a transistor (especially the small voltage difference between emitter and base). Such VOMs also have a very low voltage drop (about 50 mV) in the current ranges.

Analyzing transistor voltages. Figure 3-2 shows the basic connections for both *pnp* and *npn* transistor circuits. The coupling or bypass capacitors have been omitted to simplify the explanation. The purpose of Fig. 3-2 is to establish *normal* transistor-voltage relationships. With a normal pattern established, it is relatively simple to find an abnormal condition.

In practically all transistor circuits, the emitter-base junction must be forward-biased to get current flow through a transistor. In a *pnp* transistor, this means that the base must be made more negative (or less positive) than the emitter. Under these conditions, the emitter-base junction will draw current and cause heavy electron flow from the collector to the emitter. In an *npn* transistor, the base must be made more positive (or less negative) than the emitter in order for current to flow from emitter to collector.

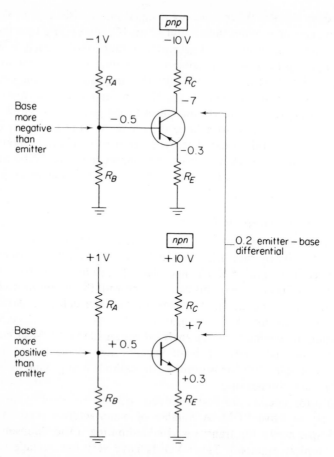

FIGURE 3-2   Basic connections for *pnp* and *npn* transistor circuits (with normal voltage relationships)

The following general rules are helpful when analyzing transistor voltage as part of troubleshooting:

1. The middle letter in *pnp* and *npn* always applies to the *base*.

2. The first two letters in *pnp* and *npn* refer to the *relative* bias polarities of the *emitter* with respect to either the base or the collector. For example, the letters *pn* (in *pnp*) indicate that the emitter is positive with respect to both the base and the collector. The letters *np* (*npn*) indicate that the emitter is negative with respect to both the base and the collector.

3. The collector-base junction is always reverse-biased.

4. The emitter-base junction is usually forward-biased. An exception is a class C amplifier (used in RF circuits).

5. A *base-input* voltage that opposes or decreases the forward bias also decreases the emitter and collector currents.

6. A *base-input* voltage that aids or increases the forward bias also increases the emitter and collector currents.

7. The dc electron flow is always against the direction of the arrow on the emitter.

8. If electron flow is into the emitter, electron flow is out from the collector.

9. If electron flow is out from the emitter, electron flow is into the collector.

Using these rules, normal transistor voltages can be summed up this way:

1. For an *npn* transistor the base is positive the emitter is not quite so positive and the collector is far more positive.

2. For a *pnp* transistor, the base is negative, the emitter is not quite so negative, and the collector is far more negative.

Measurement of transistor voltages. There are two schools of thought on how to measure transistor voltages in troubleshooting.

Element to element. Some troubleshooters prefer to measure transistor voltages from element to element (between electrodes) and note the *difference in voltage*. For example, in the circuit shown in Fig. 3-2 a 0.2 V differential exists between base and emitter. The element-to-element method of measuring transistor voltages quickly establishes forward- and reverse-bias conditions.

Element to ground. The most common method of measuring transistor voltages is to measure from a common or ground to the element. Service literature usually specifies transistor voltages this way. For example, all the voltages for the *pnp* shown in Fig. 3-2 are negative with respect to ground. (The positive test lead of the meter must be connected to ground, and the negative test lead is connected to the each of elements in turn.)

This method of labeling transistor voltages is sometimes confusing to those not familar with transistors because it appears to break the rules. (In a *pnp* transistor, some elements should be positive, but all elements are negative). However, the rules still apply.

In the case of the *pnp* shown in Fig. 3-2 the emitter is at −0.3 V; whereas the base is at −0.5. The base is *more negative* than the emitter. Thus, the emitter is *positive with respect to the base*, and the base-emitter junction is *forward-biased* (normal).

On the other hand, the base is at −0.5 V; whereas the collector is at −7 V. The base is *less negative* than the collector. Thus, the base is positive with respect to the collector, and the base-collector junction is reverse-biased (normal).

3-1.3 Troubleshooting with Transistor Voltages

This section presents an example of how voltages measured at the elements of a transistor can be used to analyze failure in solid-state circuits.

Assume that an *npn* transistor circuit is measured and that the voltages found are similar to those shown in Fig. 3-3. Except in one case, these voltages indicate a defect. It is obvious that the transistor is not forward-biased because the base is less positive than the emitter (reverse bias for an *npn*). The only circuit where this might be normal is one that requires a large *trigger* voltage or pulse (positive in this case) to turn it on.

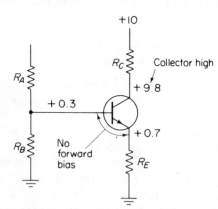

FIGURE 3-3 *npn* transistor circuit with abnormal voltages (emitter base not forward-biased, collector voltage high)

The first troubleshooting clue in Fig. 3-3 is that the collector voltage is almost as large as the collector source (at R_C). This means that very little current is flowing through R_C in the collector-emitter circuit. The transistor could be defective. However, the trouble is more likely caused by a problem in bias. The emitter voltage depends mostly on the current through R_E. Therefore, unless the value of R_E has changed substantially (this would be unusual), the problem is one of incorrect bias on the base.

The next step in this case is to measure the bias-source voltage at R_A. If the bias-source voltage is (as shown in Fig. 3-4) at 0.7 V instead of the required 2 V, the problem is obvious; the external bias voltage is incorrect. This condition will probably show up as a defect in the power supply and will appear as an incorrect voltage in other circuits.

If the source voltage is correct, as shown in Fig. 3-5, the cause of the trouble is probably a defective R_A or R_B or a defect in the transistor.

The next step is to remove all voltage from the equipment and measure the resistance of R_A and R_B. If either value is incorrect the corresponding resistor must be replaced. If both values are correct, it is reasonable to check

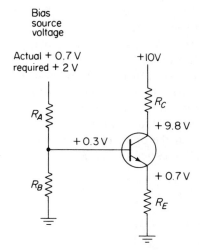

FIGURE 3-4 *npn* transistor circuit with abnormal voltages (fault traced to incorrect bias source, bias voltage low)

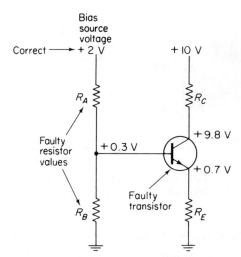

FIGURE 3-5 *npn* transistor circuit with abnormal voltages (fault traced to bias resistors or transistor)

the value of R_E. However, it is more likely that the transistor is defective. This can be established by test and/or replacement.

Practical in-circuit resistance measurements. Do not attempt to measure resistance values in transistor circuits with the resistors still connected. Although this practice may be correct for vacuum-tube circuits it is incorrect for transistor circuits. One reason is that the voltage produced by the ohm-meter battery could damage some transistors.

Even if the voltages are not dangerous the chance for an error is greater than with a transistor circuit because the transistor junctions will pass current in one direction. This can complete a circuit through other resistors and produce a series or parallel combination, thus making false indications.

This can be prevented by *disconnecting one resistor lead* before making the resistance measurement.

For example assume that an ohmmeter is connected across R_B (Figs. 3-3 to 3-5) with the negative battery terminal of the ohmmeter connected to ground, as shown in Fig. 3-6. Because R_E is also connected to ground, the negative battery terminal is connected to the end of R_E. Because the positive battery terminal is connected to the transistor base, the base-emitter junction is forward-biased, and there is electron flow. In effect, R_E is now *in parallel* with R_B, and the ohmmeter reading is incorrect. This can be prevented by disconnecting either end of R_B before making the measurement.

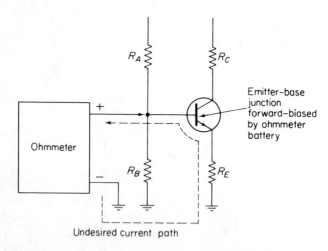

FIGURE 3-6   Example of in-circuit resistance measurements showing undesired current path through forward-biased transistor junction

3-1.4 Testing Transistors In Circuit (Forward-Bias Method)

Germanium transistors normally have a *voltage differential* of 0.2 to 0.4 V between emitter and base; silicon transistors normally have a voltage differential of 0.4 to 0.8 V. The polarities of voltages at the emitter and base depend upon the type of transistor (*npn* or *pnp*).

The voltage differential between emitter and base acts as a forward bias for the transistor. That is, a sufficient differential or forward bias will turn the transistor on, resulting in a corresponding amount of emitter-collector flow. Removal of the voltage differential or an insufficient differential will produce the opposite results. That is, the transistor is cut off (no emitter-collector flow or very little flow).

These forward-bias characteristics can be used to troubleshoot transistor circuits without removing the transistor and without using an in-circuit tester. The following sections describe two methods of testing transistors in circuit: one by removing the forward bias and the other by introducing a forward bias.

Removal of forward bias. Figure 3-7 shows the test connections for an in-circuit transistor test by removal of forward bias. The procedure is simple. First, measure the emitter-collector differential voltage under normal circuit conditions. Then, short the emitter-base junction, and note any change in emitter-collector differential. If the transistor is operating, the removal of forward bias causes the emitter-collector current flow to stop, and the emitter-collector voltage differential increases. That is, the collector voltage rises to or near the power supply value.

For example, assume that the power supply voltage is 10 V and that the differential between the collector and emitter is 5 V when the transistor is operating normally (no short between emitter and base). When the emitter-

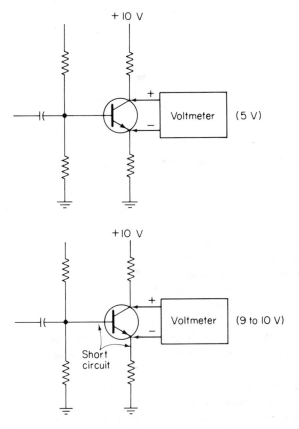

FIGURE 3-7   In-circuit transistor test (removal of forward bias)

base junction is shorted, the emitter-collector differential should rise to about 10 V (probably somewhere between 9 and 10 V).

Application of forward bias. Figure 3-8 shows the test connection for an in-circuit transistor test by the application of forward bias. The procedure is equally simple. First, measure the emitter-collector differential under normal circuit conditions. As an alternate, measure the voltage across R_E, as shown in Fig. 3-8.

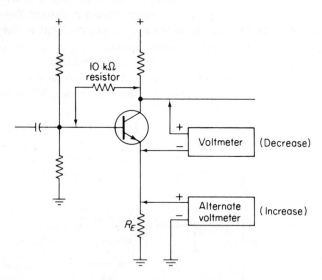

FIGURE 3-8   In-circuit transistor test (application of forward bias)

Next, connect a 10 kilohm (kΩ) resistor between the collector and base, as shown, and note any change in emitter-collector differential (or any change in voltage across R_E). If the transistor is operating the application of forward bias will cause the emitter-collector current flow to start (or increase), and the emitter-collector voltage differential will decrease, or the voltage across R_E will increase.

Go/no-go test characteristics. The test methods shown in Figs. 3-7 and 3-8 show that the transistor is operating on a go/no-go basis. This is usually sufficient for most dc and low-frequency ac applications. However, the tests do not show transistor gain or leakage. Also, the tests do not establish operation of the transistor at high frequencies or show how much delay is introduced by the transistor.

The fact that these (or similar) in-circuit tests of a transistor do not establish all the operating characteristics, particularly those that affect high-frequency RF or rapid pulse-switching applications, raises a problem in troubleshooting. Some troubleshooters reason that the only satisfactory test of a transistor is in-circuit operation. If a transistor will not perform its function in a given circuit, the transistor must be replaced. Thus, the most logical method of test is replacement.

This reasoning is generally sound, but there is one exception. It is possible that a replacement transistor will not perform satisfactorily in critical circuits (high-frequency RF and high-speed switching circuits), even though the transistor is the correct type (and may even work in another circuit). This can be misleading in troubleshooting. If such a replacement transistor does not restore the circuit to normal, the apparent fault is with another circuit, whereas the true cause of trouble is the new transistor. Fortunately, this does not happen often, even in critical circuits. However, you should be aware of the possibility.

3-1.5 Using Transistor Testers in Troubleshooting

Transistors can be tested in or out of circuit using commercial transistor testers. These testers are the solid-state equivalent of vacuum-tube testers (although they do not operate on the same principle).

The use of transistor testers in solid-state troubleshooting is generally a matter of opinion. At best, such testers show the gain and leakage of transistors at direct current or low frequencies under one set of conditions (fixed voltage, current, and so forth). For this reason, the use of transistor testers in practical troubleshooting is generally limited to home entertainment equipment. Transistors used in high-frequency or switching applications are tested by substitution (in circuit) or with special test equipment (out of circuit).

Typical commercial transistor tester. It is impractical to discuss every type of transistor tester circuit available. One very effective method for both in-circuit and out-of-circuit tests is shown in Fig. 3-9. With this method, a 60 Hz square-wave pulse is applied simultaneously to the base-emitter and the collector-emitter junctions. The current flow in each of the two junctions is measured and compared. The *difference in current flow* (collector-emitter divided by base-emitter) is the transistor gain.

It is not necessary to remove the transistor from the circuit to make such a test, but the equipment must be turned off. In fact, this test method can also show up defects in the transistor circuit.

For example, in many transistor circuits, the overall gain is set by circuit-resistance values rather than by transistor gain. Assume that a particular circuit has resistance values that would normally show a gain of 10 and that the in-circuit square-wave test (using a tester) shows no gain or very low gain. This could be the result of a bad transistor or circuit problems, or both. If the transistor is then tested out of circuit under identical conditions and a gain is shown, the problem is most likely one of an undesired change in circuit-resistance values.

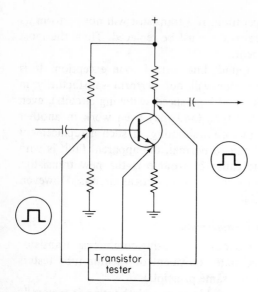

FIGURE 3-9   In-circuit transistor test method

3-1.6 Testing Transistors Out of Circuit

There are four basic tests required for transistors in practical trouble-shooting: gain, leakage, breakdown, and switching time. All these tests are best made with an oscilloscope using appropriate adapters (curve tracers, switching-characteristic checkers, and so forth). However, it is possible to test a transistor with an ohmmeter. These simple tests will show whether the transistor has leakage and whether the transistor shows some gain.

As discussed, the only true test of a transistor is in the circuit with which the transistor is to be used.

Testing transistor leakage with an ohmmeter. For test purposes (using an ohmmeter), transistors can be considered to be the same as two diodes connected back to back. Thus, each diode should show low forward resistance and high reverse resistance. These resistances can be measured with an ohmmeter, as shown in Fig. 3-10.

The same ohmmeter range should be used for each pair of measurements (base to emitter, base to collector, and collector to emitter). However, avoid using the $R \times 1$ range or an ohmmeter with a high internal battery voltage. Either of these conditions can damage a low-power transistor.

If the reverse reading is low but not shorted, the transistor is leaking.

If both forward and reverse readings are very low or show a short, the transistor is shorted.

If both forward and reverse readings are very high, the transistor is open.

If the forward and reverse readings are the same or nearly equal, the transistor is defective.

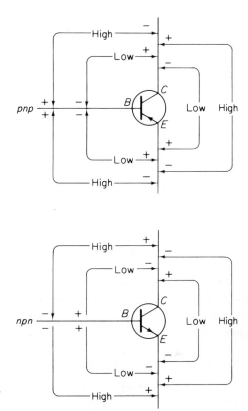

FIGURE 3-10 Transistor-leakage tests with ohmmeter

A typical forward resistance is 300 to 700 Ω. However, a low-power transistor might show only a few ohms in the forward direction, especially at the collector-emitter junction. Typical reverse resistances are 10 to 70 kΩ.

The actual resistance values depend upon the ohmmeter range and battery voltage. Thus, the *ratio of forward-to-reverse resistance* is the best indicator. Almost any transistor will show a ratio of at least 30:1. Many transistors show ratios of 100:1 or greater.

Testing transistor gain with an ohmmeter. Normally, there will be little or no current flow between emitter and collector until the base-emitter junction is forward-biased. Thus, a basic gain test of a transistor can be made using an ohmmeter. The test circuit is shown in Fig. 3-11. In this test, the $R \times 1$ range should be used. Any internal battery voltage can be used, provided that it does not exceed the maximum collector-emitter breakdown voltage.

In position A of switch S_1, there is no voltage applied to the base, and the base-emitter junction is not forward-biased. Thus, the ohmmeter reading should show a high resistance. When switch S_1 is set to B, the base-emitter

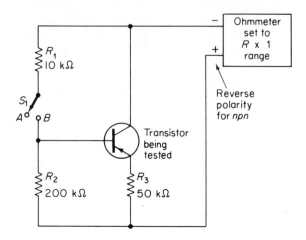

FIGURE 3-11   Transistor-gain test with ohmmeter

circuit is forward-biased (by the voltage across R_1 and R_2), and current flows in the emitter-collector circuit. This is indicated by a lower resistance reading on the ohmmeter. A 10:1 resistance ratio is typical for an AF transistor.

3-1.7 Testing Diodes Out of Circuit

There are three basic tests required for diodes. First, the diode must have the ability to pass current in one direction (forward current) and prevent or limit current flow (reverse current) in the opposite direction. Second, for a given reverse voltage the reverse current should not exceed a given value. Third, for a given forward current, the voltage drop across the diode should not exceed a given value. If a diode is to be used in pulse or digital work, the switching time must also be tested. These tests are best performed using a scope with appropriate adapters.

However, because the elementary purpose of a diode is to prevent current flow in one direction while passing current in the opposite direction, a diode can be tested using an ohmmeter for basic troubleshooting purposes. In this case, the ohmmeter is used to measure *forward* and *reverse resistance* of the diode. The basic circuit is shown in Fig. 3-12.

A good diode will show high resistance in the reverse direction and low resistance in the forward direction.

If resistance is low in the reverse direction, the diode is probably leaking.

If resistance is high in both directions, the diode is probably open.

A low resistance in both directions usually indicates a shorted diode.

It is possible for a defective diode to show a difference in forward and reverse resistance. The important factor in making a diode-resistance test

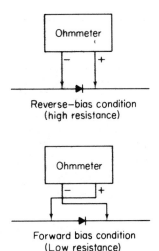

Reverse–bias condition
(high resistance)

Ohmmeter

FIGURE 3-12 Basic diode test
with ohmmeter

Forward bias condition
(Low resistance)

is the *ratio of forward-to-reverse resistance* (often known as the *front-to-back ratio* or the *back-to-front ratio*). The actual ratio depends upon the type of diode. However, as a guideline, a small-signal diode has a ratio of several hundred to one; whereas a power rectifier can operate satisfactorily with a ratio of 10: 1.

3-1.8 Troubleshooting ICs

There is some difference of opinion on testing ICs in circuit or out of circuit during troubleshooting. An in-circuit test is the most convenient because the power source is available and you do not have to unsolder the IC. (Removal and replacement of an IC can be quite a job.)

Of course, first you must measure the dc voltages applied at the IC terminals to make sure that they are available and correct. If the voltages are absent or abnormal, this is a good starting point for troubleshooting.

With the power sources established, the in-circuit IC is tested by applying the appropriate input and monitoring the output. In some cases, it is not necessary to inject an input because the normal input is supplied by the circuits ahead of the IC.

One drawback to testing an IC in circuit is that the circuits before (input) and after (output) the IC may be defective. This can lead you to think that the IC is bad. For example, assume that the IC is used as the IF stages of a radio receiver. To test such an IC, you inject a signal at the IC input and monitor the IC output. Now, assume that the IC output terminal is connected to a short circuit. There will be no output indicated, even though the IC and the input signals are good. Of course, this will show up as an incorrect resistance measurement (if such measurements are made).

Out-of-circuit tests for ICs have two obvious disadvantages: You must remove the IC, and you must supply the required power. However, if you test a suspected IC after removal and find that it is operating properly out of circuit, it is logical to assume that there is trouble in the circuits connected to the IC. This is very convenient to know before you go to the trouble of installing a replacement IC.

IC voltage measurements. Although the test procedures for an IC are the same as those used for conventional transistor circuits of the same type, measurement of the static (dc) voltage applied to the IC is not identical. Most ICs require connection to both a positive and a negative power source, but a few ICs can be operated from a single power supply source.

Many ICs require equal power supply voltages (such as $+9$ V and -9 V). However, this is not the case with the circuit shown in Fig. 3-13, which requires $+9$ V at pin 8, and -4.8 V at pin 4.

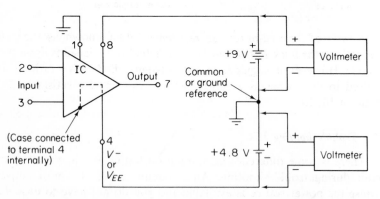

FIGURE 3-13 Measuring static (power source) voltages of ICs during troubleshooting

In the case of most transistor circuits, it is common to label one power supply lead positive and the other negative without specifying which (if either) is common or ground. But in the case of ICs, it is necessary that *all* IC power supply voltages be referenced to a common or ground.

Manufacturers do not agree on power supply labeling for ICs. For example, one manufacturer might use V+ to indicate the positive voltage and V− to indicate the negative voltage. Another might use the symbols V_{EE} and V_{CC} to represent negative and positive, respectively. For this reason, you should study the schematic diagram carefully before measuring power source voltages during troubleshooting.

No matter what labeling is used, the IC will require two power sources, with the positive lead of one and the negative lead of the other tied to ground. Each voltage must be measured separately, as shown in Fig. 3-13.

Note that the IC case (such as a TO-5 type) of the circuit shown in Fig. 3-13 is connected to pin 4. Such a connection is typical for most ICs (but not necessarily pin 4). Thus, the case will be below ground (or hot) by 4.8 V.

3-1.9 Effects of Capacitors in Troubleshooting

During the troubleshooting process, suspected capacitors can be removed from the circuit and tested on the bridge-type capacitor checkers discussed in Sec. 2-4. This will establish that the capacitor value is correct. If the checker shows the value to be correct, it is reasonable to assume that the capacitor is not open, shorted, or leaking.

From another standpoint, if a capacitor shows no shorts, opens, or leakage, it is also reasonable to assume that the capacitor is good. Thus, from a *practical* troubleshooting standpoint, a simple test that shows the possibility of shorts, opens, or leakage is usually sufficient.

There are two basic methods for a quick check of capacitors during troubleshooting. One method involves using the circuit voltages. The other technique requires an ohmmeter.

Checking capacitors with circuit voltages. As shown in Fig. 3-14(a), this method involves disconnecting one lead of the capacitor (the ground or cold lead) and connecting a voltmeter between the disconnected lead and ground. In a good capacitor, there should be a momentary voltage indication (or surge) as the capacitor charges up the voltage at the hot end.

If the voltage indication remains high, the capacitor is probably shorted.

If the voltage indication is steady but not necessarily high, the capacitor is probably leaking.

If there is no voltage indication whatsoever, the capacitor is probably open.

Checking capacitors with an ohmmeter. As shown in Fig. 3-14(b), this method involves disconnecting one lead of the capacitor (usually the hot end) and connecting an ohmmeter across the capacitor. Make certain all power is removed from the circuit. As a precaution, short across the capacitor to make sure that no charge is being retained after the power is removed. In a good capacitor, there should be a momentary resistance indication (or surge) as the capacitor charges up to the voltage of the ohmmeter battery.

If the resistance indication is near zero and remains so, the capacitor is probably shorted.

If the resistance indication is steady at some high value, the capacitor is probably leaking.

If there is no resistance indication whatsoever, the capacitor is probably open.

Functions of capacitors in circuits. The functions of capacitors in solid-state circuits are similar to the functions of those in vacuum-tube equip-

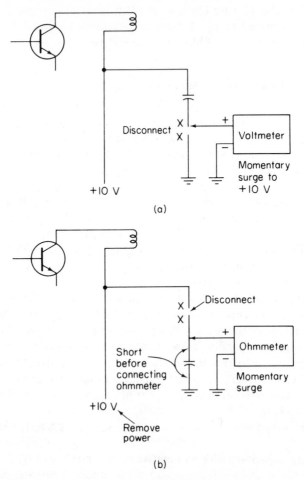

FIGURE 3-14   Check capacitors with circuit voltages (power applied) and with ohmmeter (power removed)

ment. However, the results produced by capacitor failure are not necessarily the same. An emitter-bypass capacitor is a good example.

The emitter resistor in a solid-state circuit (such as R_4 in Fig. 3-15) is used to stabilize the transistor's dc gain and prevent thermal runaway. With an emitter resistor in the circuit, any increase in collector current produces a greater drop in voltage across the resistor. When all other factors remain the same, the change in emitter voltage reduces the base-emitter forward-bias differential, thus tending to reduce collector current flow. When circuit stability is more important than gain, the emitter resistor is not bypassed. When ac or signal gain must be high, the emitter resistance is bypassed to permit passage of the signal. If the emitter-bypass capacitor is open, stage

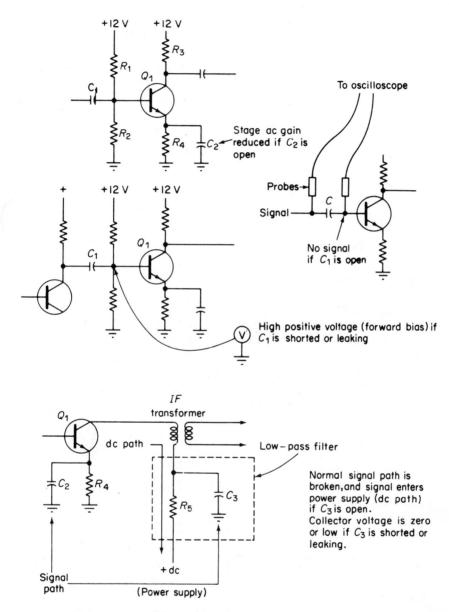

FIGURE 3-15 Effects of capacitor failure in solid-state circuits

gain is reduced drastically, although the transistor's dc voltages remain substantially the same.

Low-gain symptoms. If there is a low-gain symptom in any solid-state amplifier with an emitter bypass and the voltages appear normal, check the bypass capacitor. This can be done by shunting the bypass with a known-good capacitor of the same value. As a precaution, shut off the power before connecting the shunt capacitor; then, reapply power. This will prevent damage to the transistor (caused by large current surges).

Coupling capacitors. The functions of coupling (and decoupling) capacitors in solid-state circuits are essentially the same as the functions of those in vacuum-tube equipment. However, the capacitance values are much larger for solid-state circuits, particularly at *low* frequencies. Electrolytic capacitors are usually required in solid-state circuit to get the large capacitance values. From a troubleshooting standpoint, electrolytics tend to have more leakage than mica or ceramic capacitors. However, good-quality electrolytics (typically the bantam type found in solid-state circuits) have leakage of less than 10 μA at normal operating voltage.

Defects in coupling capacitors. The function of C_1 in Fig. 3-15 is to pass signals from the previous stage to the base of Q_1. If C_1 is shorted or leaking badly, the voltage from the previous stage is applied to Q_1. This forward-biases Q_1, causing heavy current flow and possible burnout of the transistor. In any event, Q_1 is driven into saturation, and stage gain is reduced.

If C_1 is open, there is little or no change in the voltage at Q_1, but the signal from the previous stage will not appear at the base of Q_1. From a troubleshooting standpoint, a shorted or leaking C_1 will show up as abnormal voltages (and probably as distortion of the signal wave form). If C_1 is suspected of being shorted or leaky, replace it. An open C_1 will show up as a lack of signal at the base of Q_1, with a normal signal at the previous stage. If an open C_1 is suspected, replace C_1 or try shunting it with a known-good capacitor, whichever is convenient.

Defects in decoupling or bypass capacitors. The function of C_3 in Fig. 3-15 is to pass operating-signal frequencies to ground (to provide a return path) and to prevent signals from entering the power supply line or other circuits connected to the line. In effect, C_3 and R_5 form a low-pass filter that passes dc and very low frequency signals (well below the operating frequency of the circuit) through the power supply line. Higher-frequency signals are passed to ground and do not enter the power supply line.

If C_3 is shorted or leaking badly, the power supply voltage will be shorted to ground or greatly reduced. This reduction of collector voltage will make the stage totally inoperative or will reduce the output, depending on the amount of leakage in C_3.

If C_3 is open, there will be little or no change in the voltages at Q_1. However, the signals will appear in the power supply line. Also, signal gain will be reduced, and the signal wave form will be distorted. In some cases, at higher signal frequencies, the signal simply cannot pass through the power supply circuits. Because there is no path through an open C_3, the signal will not appear on the collector circuit. From a practical troubleshooting standpoint, the results of an open C_3 depend upon the values of R_s (and the power supply components) as well as on the signal frequency involved.

3-1.10 Effects of Voltage on Circuit Resistance

The effects of shorts on resistors are less drastic in solid-state circuits than in vacuum-tube circuits because of the lower voltage used in solid-state circuits. For example, except for a few high-power devices, most solid-state circuits operate at voltages well below 25 V. Typically, solid-state power sources are 12 V or less. A 1 kΩ resistance shorted directly across a 25 V source produces only 25 mA current flow, or about 0.6 W. A 1 W resistor can easily handle this power with no trouble. Even a 0.5 W resistor will probably survive a temporary short of this level.

On the other hand, the same resistance across a 300 V source (typical for vacuum-tube circuits) produces about 0.3 ampere (A) current flow, or about 90 W. This will destroy all but heavy power resistors.

For these reasons, resistors do not burn out as often in solid-state equipment as they do in similar vacuum-tube equipment. Similarly, solid-state resistance values do not usually change as a result of prolonged heating. There are exceptions, of course, but most solid-state troubles are the result of defects in capacitors, transistors, and diodes, in that order.

3-1.11 Effects of Voltage on Poor Solder Joints

The low voltages in solid-state equipment have just the opposite effect on poor solder joints (so-called cold solder joints) and partial breaks in printed wiring. The high voltages in vacuum-tube equipment can often overcome the resistance created by cold solder joints and partial printed-circuit breaks.

When there is no obvious cause for a low voltage at some point in the circuit or there is an abnormally high resistance, look for cold solder joints or defects in printed-circuit wiring. Use a magnifying glass to locate defects in printed wiring. *Minor* breaks in printed wiring can sometimes be repaired by applying solder at the break. However, this is recommended only as a temporary measure. Under emergency conditions, it is possible to run a wire between two points on either side of the break. However, it is recommended that the entire board be replaced as soon as practical.

Finding cold solder joints. Cold solder joints can sometimes be found with an ohmmeter. Remove all power. Connect the ohmmeter across two wires leading out of the suspected cold solder joint, as shown in Fig. 3-16. Flex the wires by applying pressure with the ohmmeter prod tips. Switch the ohmmeter to different ranges, and check for any change in resistance. For example, a cold solder joint can appear to be good on the high ohmmeter ranges but an open on the lower ranges. Look for resistance indications that

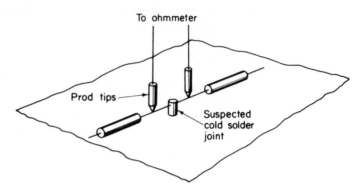

FIGURE 3-16   Locating cold solder joints with ohmmeter

tend to drift or change when the ohmmeter is returned to a particular scale. If a cold solder joint is suspected, reheat the joint with a soldering tool; then, recheck the resistance.

3-2. AMPLIFIER TROUBLESHOOTING

Before presenting a step-by-step example of amplifier troubleshooting, we shall discuss the basic troubleshooting approach, the test procedures normally associated with amplifier troubleshooting, and some practical notes on analysis of basic amplifier circuit.

3-2.1 Basic Amplifier Troubleshooting Approach

The basic troubleshooting approach for an amplifier involves *signal tracing*. The input and output wave forms of each stage can be monitored on an oscilloscope or voltmeter. Any stage showing an abnormal wave form (in amplitude, wave shape, and so forth) or the absence of an output wave form with a known-good input signal points to a defect in that stage. Voltage and resistance measurements on all elements of the vacuum tube or transistor will then pinpoint the problem.

An oscilloscope is the most logical instrument for checking amplifier circuits, whether they are complete amplifier systems or a single stage. The scope can duplicate every function of an electronic voltmeter in troubleshooting. In addition, the scope offers the advantage of a visual display for common audio-amplifier conditions such as distortion, hum, noise, ripple, and oscillation.

When you are troubleshooting amplifier circuits using signal tracing, a scope is used in much the same manner as a voltmeter. A signal is introduced into the input by a signal generator, as shown in Fig. 3-17. The

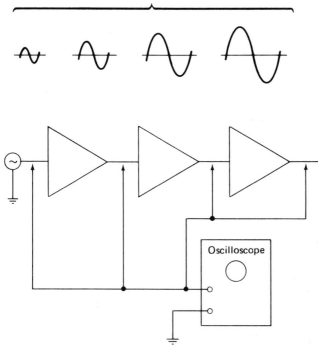

FIGURE 3-17   Basic signal tracing through amplifier circuits using sine waves and an oscilloscope

amplitude and wave form of the input signal are measured on the scope. The scope probe is then moved to the input and output of each stage, in turn, until the final output is reached. The gain of each stage is measured as a voltage on the scope. In addition, it is possible to observe any change in wave form from that applied to the input. Thus, stage gain and distortion (if any) are established quickly with a scope.

3-2.2 Amplifier Frequency Response

The frequency response of an audio amplifier can be measured with an audio-signal generator and a meter or scope. When a meter is used, the signal generator is tuned to various frequencies, and the resultant circuit-output response is measured at each frequency. The results are then plotted in the form of a graph or *response curve*, as shown in Fig. 3-18.

The procedure is essentially the same when a scope is used to measure audio-circuit frequency response. However, the scope gives the added benefit of a visual analysis of distortion, as discussed later in this chapter.

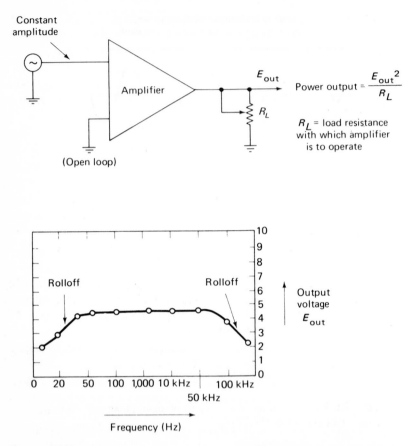

FIGURE 3-18 Amplifier frequency-response test connections and typical response curve

The basic procedure for measurement of frequency response (with either meter or scope) is to apply a *constant-amplitude signal* while monitoring the circuit output. The input signal is varied in frequency (but not in amplitude) across the entire operating range of the circuit. Any well-designed audio circuit should have a constant response from about 20 Hz to 20 kHz. With direct-coupled amplifiers, the response is usually extended from a few hertz (or possibly from direct current) up to 100 kHz (and higher). The voltage output at various frequencies across the range is plotted on a graph as follows:

1. Connect the equipment, as shown in Fig. 3-18.

2. Initially, set the generator-output frequency to the low end of the range. Then, set the generator-output amplitude to the desired input level.

3. In the absence of a realistic test-input voltage, set the generator output to an arbitrary value. A simple method of finding a satisfactory input level is to monitor the circuit output (with the meter or scope) and increase the generator output at the circuit's center frequency (or at 1 kHz) until the circuit is overdriven. This point is indicated when further increases in the generator output do not cause further increases in meter reading (or the output wave-form peaks begin to flatten on the scope display). Set the generator output *just below* this point. Then, return the meter or scope to monitor the generator voltage (at the circuit input) and measure the voltage. Keep the generator at this voltage *throughout* the test.

4. If the circuit is provided with any operating or adjustment controls (volume, loudness, gain, treble, balance, and so forth), set these controls to some arbitrary point when making the initial frequency-response measurement. The response measurements can then be repeated at different control settings if desired.

5. Record the circuit-output voltage on the graph. Without changing the generator-output amplitude, increase the generator frequency by some fixed amount, and record the new circuit-output voltage. The amount of frequency increase between each measurement is an arbitrary matter. Use an increase of 10 Hz where rolloff occurs and 100 Hz at the middle frequencies.

6. Repeat the process, checking and recording the amplifier-output voltage at each of the checkpoints in order to obtain a frequency-response curve. For a typical audio amplifier, the curve will resemble that shown in Fig. 3-18, with a flat portion across the middle frequencies and a rolloff at each end.

7. After the initial frequency-response check, the effects of operating or adjustment controls should be checked. Volume, loudness, and gain controls should have the same effect all across the frequency range. Treble and bass controls may also have some effect at all frequencies. However, a treble control should have the greatest effect at the high end; whereas a bass control should have the greatest effect at the low end.

8. Note that generator output *may* vary with changes in frequency, a fact often overlooked in making a frequency-response test during troubleshooting. Even precision laboratory generators can vary in output with changes in frequency, thus resulting in considerable error. It is recommended that the generator output be monitored after each change in frequency (some audio generators have a built-in output meter). Then, if necessary, the generator-output amplitude can be reset to the correct value. It is more important that the generator-output amplitude remain *constant* rather than set at some specific value when making a frequency-response check.

3-2.3 Amplifier Voltage-Gain Measurement

Voltage gain in an audio amplifier is measured in the same way as frequency response. The ratio of output voltage to input voltage (at any given frequency or across the entire frequency range) is the voltage gain. Because the input voltage (generator output) is held constant for a frequency-response test, a voltage-gain curve should be identical to a frequency-response curve.

3-2.4 Power-Output and -Gain Measurement

The power output of an audio amplifier is found by noting the output voltage E_{out} across the load resistance R_L (Fig. 3-18), at any frequency or across the entire frequency range. Power output is E_{out}^2/R_L.

To find power gain of an amplifier, it is necessary to find both the input and the output power. Input power is found in the same way as output power except that the input impedance must be known (or calculated). Calculating input impedance is not always practical in the case of some amplifiers, especially in designs where input impedance is dependent upon transistor gain. (The procedure for finding input impedance of an amplifier is described in Sec. 3-2.8). With input power known (or estimated), the power gain is the ratio of output power to input power.

In some applications, the input-sensitivity specification is used. Input-sensitivity specifications require a minimum power output with a given voltage input (such as 100 W output with a 1 V rms input).

3-2.5 Power-Bandwidth Measurement

Many audio-amplifier design specifications include a power-bandwidth factor. Such specifications require that the audio amplifier deliver a given power output across a given frequency range. For example, a circuit may produce full power output up to 20 kHz, even though the frequency response is flat up to 100 kHz. That is, voltage (without load) remains constant up to 100 kHz; whereas power output (across a normal load) remains constant up to 20 kHz.

3-2.6 Load-Sensitivity Measurement

An audio-amplifier circuit of any design, especially power amplifiers, is sensitive to changes in load. An amplifier produces maximum power when the output impedance is the same as the load impedance.

The circuit for load-sensitivity measurement is the same as the circuit for frequency response (Fig. 3-18) except that load resistance R_L is variable. (Never use a wirewound load resistance. The reactance can result in considerable error.)

Measure the power output at various load-impedance versus output-impedance ratios. That is, set R_L to various resistance values, include a value equal to the amplifier-output impedance, and then note the voltage and/or power gain at each setting. Then, repeat the test at various frequencies. Figure 3-19 shows a typical load-sensitivity response curve. Note that if the load is twice the output impedance (as indicated by a 2:0 ratio in Fig. 3-19), the output power is reduced to approximately 50 percent.

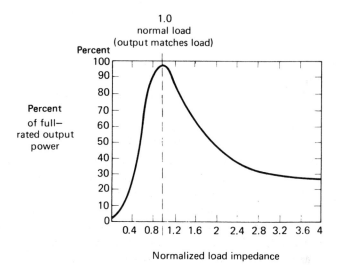

FIGURE 3-19   Output power versus load impedance (showing effect of match and mismatch between output and load)

3-2.7 Dynamic Output-Impedance Measurement

The load-sensitivity test can be reversed to find the dynamic output impedance of an amplifier circuit. The connections (Fig. 3-18) and procedures are the same except that R_L is varied until *maximum* output power is found. Power is removed, and R_L is disconnected from the circuit. The dc resistance of R_L is equal to the dynamic output impedance. Of course, the value applies only at the frequency of measurement. The test can be repeated across the entire frequency range if desired.

3-2.8 Dynamic Input-Impedance Measurement

To find the dynamic input impedance of an amplifier, use the circuit shown in Fig. 3-20. The test conditions are identical to those for frequency response, power output, and so on. Move switch S between points A and B while adjusting resistance R until the voltage reading is the same in *both* positions of S. Disconnect R, and measure the dc resistance of R, which is then equal to the dynamic impedance of the amplifier input.

Accuracy of the impedance measurement is dependent upon the accuracy with which the dc resistance is measured. A noninductive (not wire-

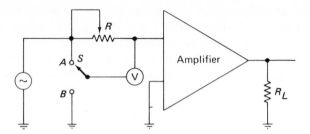

FIGURE 3-20   Amplifier dynamic input-impedance test connections

wound) resistance must be used. The impedance found by this method applies only to the frequency used during the test.

3-2.9 Checking Distortion by Sine-Wave Analysis

All amplifiers are subject to possible distortion. That is, the output signal may not be identical to the input signal. Theoretically, the output should be identical to the input except for amplitude. Some troubleshooting techniques are based on analyzing the wave shape of signals passing through an amplifier to determine possible distortion. If distortion (or an abnormal amount of distortion) is present, the circuit is then checked further by the usual troubleshooting methods (localization, voltage measurements, and so forth).

Amplifier distortion can be checked by sine-wave analysis. The procedures are the same as those used for signal tracing (Sec. 3-2.1 and Fig. 3-17). However, the primary concern in distortion analysis is deviation of the amplifier-output (or stage-output) wave form from the input wave form. If there is no change (except in amplitude), there is no distortion. If there is a change in the wave form, the nature of the change often reveals the cause of distortion. For example, the present of second or third harmonics distorts the fundamental.

In practical troubleshooting, analyzing sine waves to pinpoint amplifier problems that produce distortion is a difficult job that requires considerable experience. Unless the distortion is severe, it may pass unnoticed. Sine waves are best used where *harmonic-distortion* or *intermodulation-distortion* meters are combined with the scope for distortion analysis. If a scope is to be used alone, *square waves* provide the best basis for distortion analysis. (The reverse is true for frequency-response and power measurements.)

3-2.10 Checking Distortion by Square-Wave Analysis

The procedure for checking distortion by means of square waves is essentially the same as that used with sine waves. Distortion analysis is more effective with square waves because of their high odd-harmonic content and because it is easier to see a deviation from a straight line with sharp corners than from a curving line.

Square waves are introduced into the circuit input, and the output is monitored with a scope, as shown in Fig. 3-21. The primary concern is deviation of the amplifier-output (or stage-output) wave form from the input wave form (which is also monitored on the scope). If the scope has the *dual-trace feature*, the input and output can be monitored simultaneously. If there is a change in wave form, the nature of the change often reveals the cause of the distortion.

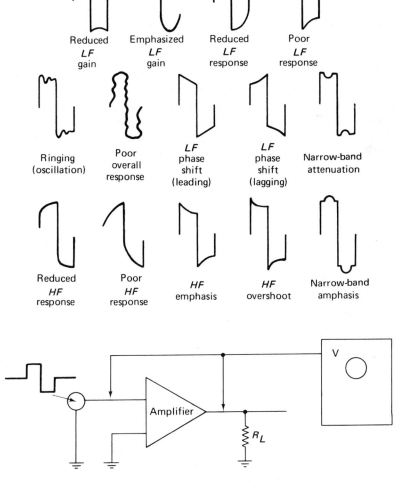

FIGURE 3-21   Amplifier square-wave distortion analysis

The third, fifth, seventh, and ninth harmonics of a clean square wave are emphasized. If an amplifier passes a given frequency and produces a clean square-wave output, it is safe to assume that the frequency response is good up to at least nine times the square-wave frequency.

3-2.11 Harmonic-Distortion Measurement

No matter what amplifier circuit is used or how well it is designed, there is always the possibility of odd or even harmonics being present with the fundamental. These harmonics combine with the fundamental and produce distortion, as is the case when any two signals are combined. The effects of second- and third-harmonic distortion are shown in Fig. 3-22.

Commercial harmonic-distortion meters operate on the *fundamental-suppression* principle. A sine wave is applied to the amplifier input, and the output is measured on the scope, as shown in Fig. 3-22. The output is then applied through a filter that suppresses the fundamental frequency. Any output from the filter is then the result of harmonics.

The output is also displayed on the scope. (Some commercial harmonic-distortion meters use a built-in meter instead of, or in addition to, an external scope.) When the scope is used, the frequency of the filter-output signal is checked to determine harmonic content. For example, if the input is 1 kHz and the output (after filtering) is 3 kHz, *third-harmonic distortion* is indicated.

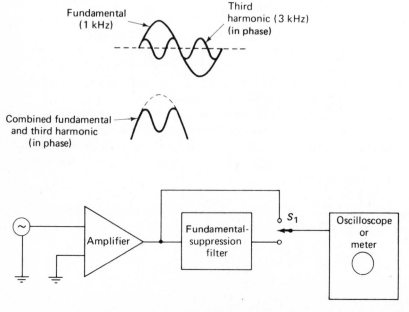

FIGURE 3-22   Harmonic-distortion measurement

The percentage of harmonic distortion is also determined by this method. For example, if the output is 100 mV without filter and 3 mV with filter, a 3 percent harmonic distortion is indicated.

On some commercial harmonic-distortion meters, the filter is tunable, so that the amplifier can be tested over a wide range of fundamental frequencies. On other harmonic-distortion meters, the filter is fixed in frequency but can be detuned slightly to produce a sharp null.

3-2.12 Intermodulation-Distortion Measurement

When two signals of different frequencies are mixed in an amplifier, there is a possibility that the lower-frequency signal will modulate the amplitude of the higher-frequency signal. This produces a form of distortion known as *intermodulation distortion*.

Commercial intermodulation-distortion meters consist of a signal generator and high-pass filter, as shown in Fig. 3-23. The signal-generator portion of the meter produces a high-frequency signal (usually about 7 kHz) that is modulated by a low-frequency signal (usually 60 Hz). The mixed signals are applied to the circuit input. The amplifier output is connected through a high-pass filter to the scope's vertical channel. The high-pass filter removes the low-frequency (60 Hz) signal. Thus, the only signal appearing on the scope's vertical channel should be the high-frequency (7 kHz) signal. If any 60 Hz signal is present on the display, it is being passed through as modulation on the 7 kHz signal.

Figure 3-23 also shows an intermodulation test circuit that can be fabricated in the shop or laboratory. Note that the high-pass filter is designed to pass signals above about 200 Hz. The purpose of the 40 and 10 kΩ resistors is to set the 60 Hz signal at four times the amplitude of the 7 kHz signal. Most audio generators provide for a line-frequency output (60 Hz) that can be used as the low-frequency modulation source.

If the laboratory circuit shown in Fig. 3-23 is used instead of a commercial meter, set the generator line-frequency (60 Hz) output to 2 V (if adjustable) or to some value that will not overdrive the amplifier being tested. Then, set the audio-generator output (7 kHz) to 2 V (or to the same value as the 60 Hz output).

Calculate the percentage of intermodulation distortion using the equation shown in Fig. 3-23. For example, if the maximum output (as shown on the scope) is 1 V and the minimum is 0.9 V, the percentage of intermodulation is approximately

$$\frac{1.0 - 0.9}{1.0 + 0.9} = 0.05$$

and

$$0.05 \times 100 = 5\%$$

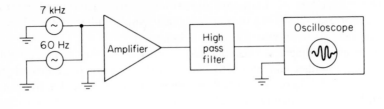

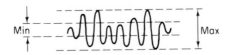

$$\% \text{ intermodulation distortion} = 100 \times \frac{max - min}{max + min}$$

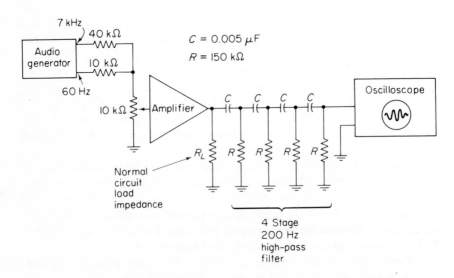

FIGURE 3-23 Intermodulation-distortion measurement

3-2.13 Background-Noise Measurement

If the vertical channel of a scope is sufficiently sensitive, the scope can be used to check and measure the background-noise level of an amplifier, as well as to check for the presence of hum, oscillation, and the like. The scope's vertical channel should be capable of a measurable deflection with about 1 mV (or less) because this is the background-noise level of many amplifiers.

The basic procedure consists of measuring amplifier output with the volume or gain control (if any) at maximum but without an input signal. The oscilloscope is superior to a voltmeter for noise-level measurement

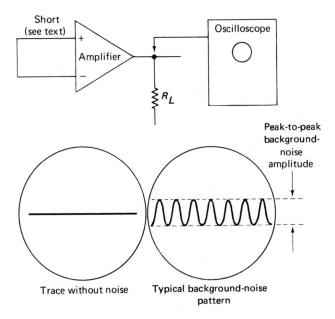

FIGURE 3-24   Amplifier background-noise test connections

because the frequency and nature of the noise (or other signal) are displayed visually.

The basic connections for measuring the level of background noise are shown in Fig. 3-24. The scope gain is increased until there is a noise or hash indication.

It is possible that a noise indication can be caused by pickup in the leads between the amplifier output and the scope. If in doubt, disconnect the leads from the amplifier but not from the scope.

If it is suspected that there is a 60 Hz line hum present in the amplifier output (picked up from the power supply or any other source), set the scope's SYNC control (or whatever other control is required to synchronize the scope trace at the line frequency) to LINE. If a stationary signal pattern appears, it is the result of the line hum.

If a signal appears that is not at the line frequency, it can be the result of oscillation in the amplifier or stray pickup. Short the amplifier's input terminals. If the signal remains, it is probably oscillation in the amplifier.

3-2.14 Feedback-Amplifier Troubleshooting

Troubleshooting amplifiers without feedback is a relatively simple procedure. When the amplifier has feedback, the task is more difficult. Problems such as measurement of gain can be of particular concern.

For example, if you try opening the loop to make gain measurements, you usually find so much gain that the amplifier saturates and the measurements are meaningless. On the other hand, if you start making wave-form measurements on a working closed-loop system, you often find the input and output signals are normal (or near normal), although many of the waveforms are distorted inside the loop. For this reason, feedback loops, especially internal-stage feedback loops, require special attention.

Typical feedback-amplifier circuits. Figure 3-25 is the schematic of a basic feedback amplifier. Note the various wave forms around the circuit. These wave forms are similar to those that appear if the amplifier is used with sine waves. Note that there is an approximate 15 percent distortion inside the feedback loop (between Q_1 and Q_2) but only 0.5 percent distortion at the output. This is only slightly greater distortion than the input 0.3 percent. Open-loop gain for this circuit is approximately 4,300; closed-loop gain is approximately 1,000. The gain ratio (open loop to closed loop) of 4:1 is typical for feedback amplifiers.

Amplification of signals. Transistors in feedback amplifiers behave just like transistors in any other circuits. That is, the transistors respond to all the same rules for gain and input-output impedance. Specifically, each transistor amplifies the signal appearing between its emitter and its base. It

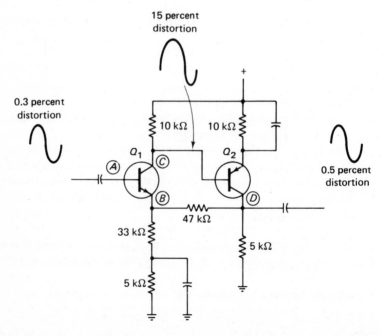

FIGURE 3-25 Basic feedback amplifier

is here that the greatest difference between gain stages in feedback amplifiers and gain stages in nonfeedback (open-loop) amplifiers occurs.

Difference in open-loop and closed-loop gain. Transistor Q_1 in Fig. 3-26 has a varying signal on both the emitter and the base rather than on one element. In a nonfeedback amplifier, the signal usually varies at only one element, either the emitter or the base. Because most feedback systems use negative feedback, the signals at both the base and the emitter are in phase. The resultant gain is much less than when one of these elements is fixed (no feedback, open loop).

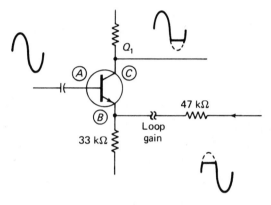

FIGURE 3-26   Amplifier-induced distortion in signal returning to point *B*

This accounts for the amplifier's great gain increase when the loop is opened. Either the base or the emitter of the transistor stops moving, and a much larger effective input signal appears at the base-emitter control element. Assume that a perfect input signal is applied to the input (point *A* in Fig. 3-26). If the amplifier is perfect (produces no distortion), the signal returning to *B* will also be undistorted. Because the system uses negative feedback, the signal that travels around the loop a second time is undistorted as well. If the amplifier is not perfect (assume an extreme case of clipping distortion), the returning signal will show that effect of distortion, as illustrated in Fig. 3-26.

To simplify the explanation, assume that the clipping is introduced in Q_1 and that Q_2 is perfect. Now, the signals applied to the base and emitter of Q_1 are not identical. The resultant applied signal at the control point of Q_1 will be quite distorted. In effect, the distortion will be a mirror image of the distortion introduced by Q_1. Transistor Q_1 then amplifies this distortion and adds its own counterdistortion. The result, then, after many trips around the loop, is that there can be distortion *inside the loop* but that this is counterbalanced by the feedback system. The final output from Q_2 is undistorted or relatively free of amplifier-induced distortion. The higher the amplification and the greater the feedback, the more effective this cancellation becomes, and the lower the output distortion becomes.

This last fact marks the basic difference in troubleshooting a feedback amplifier. In any amplifier, there are three basic causes of distortion: *overdriving*, operating the transistor at the *wrong bias point*, and the *inherent nonlinearity* of any solid-state device.

Overdriving can be the result of many causes (too much input signal, too much gain in the previous stage, and so forth). The net result is that the output signal is clipped on one peak as a consequence of the transistor being driven into saturation and on the other peak by driving the transistor below cutoff.

Operating at the *wrong bias point* can also produce clipping, but of only one peak. For example, if the input signal is 1 V and the transistor is biased at 1 V, the input will swing from 0.5 to 1.5 V. Assume that the transistor saturates at any point where the base goes above 1.6 V and is cut off when the base goes below 0.4 V. No problem occurs when the bias is correct at 1 V.

But now assume that the bias point is shifted (because of component aging, transistor leakage, and so forth) to 1.3 V. When the 1 V input signal is applied, the base swings from 0.8 to 1.8 V, and the transistor saturates when one peak goes from 1.6 to 1.8 V. If, on the other hand, the bias point is shifted down to 0.7 V, the base swings from 0.2 V to 1.2 V, and the opposite peak is clipped as the transistor goes into cutoff.

Even if the transistor is not overdriven, it is still possible to operate a transistor on a nonlinear portion of its curve because of wrong bias. Some portion of the input-output curve of all transistors is more linear than other portions. That is, the output increases (or decreases) directly in proportion to input. An increase of 10 percent at the input produces an increase of 10 percent at the output. Ideally, transistors are operated at the center of this linear curve. If the bias point is changed, the transistor can operate on a portion of the curve that is less linear than the desired point.

The *inherent nonlinearity* of any solid-state device (diode, transistor, and so forth) can produce distortion even if a stage is not overdriven and is properly biased. That is, the output never increases (or decreases) directly in proportion to the input. For example, an increase of 10 percent at the input can produce an increase of 13 percent at the output. This is one of the main reasons for feedback in amplifiers where low distortion is required.

In summary, a negative-feedback loop operates to minimize distortion, in addition to stabilizing gain. The feedback-takeoff point has the least distortion of any point within the loop. From a practical troubleshooting standpoint, if the *final* output distortion and the overall gain are within limits, all the stages within the loop can be considered to be operating properly. Even if there is some abnormal gain in one or more of the stages, the overall feedback system has compensated for the problem. Of course, if the overall gain and/or distortion are not within limits, the individual stages must be checked.

3-2.15 Feedback-Amplifier Troubleshooting Procedures

Most feedback-amplifier problems can be pinpointed by wave-form and voltage measurements, as discussed throughout this book. You should give special attention to the following paragraphs when you are troubleshooting any feedback-amplifier circuit.

Opening the loop. Some troubleshooting literature recommends that the loop be opened and the circuits checked under no-feedback conditions. In some cases, this can cause circuit damage. But even if there is no damage, the technique is rarely effective. Open-loop gain is usually so great that some stage will block or distort badly. If the technique is used, as it must be for some circuits, keep in mind that distortion is *increased* when the loop is opened. That is, a normally closed-loop amplifier can show considerable distortion when operated as an open-loop device, even though the amplifier is good.

Measuring stage gain. Care should be taken when measuring the gain of amplifier stages in a feedback amplifier. For example, in Fig. 3-25, if you measure the signal at the base of Q_1, the base-to-ground voltage is not the same as the input voltage. To get the correct value, connect the low side of the measuring device (ac voltmeter or scope) to the emitter and the other lead (high side) to the base, as shown in Fig. 3-27. In effect, measure the

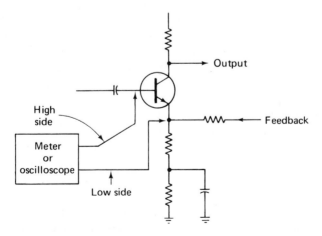

FIGURE 3-27 Measuring input-signal voltage or wave forms

signal that appears *across* the base-emitter junction. This measurement will include the effect of the feedback signal.

As a general safety precaution, *never* connect the ground lead of a voltmeter or scope to the base of a transistor unless the lead connects back to an isolated inner chassis on the meter or scope. The reason for this precaution is that large ac ground-loop currents (between the measuring device and the equipment being serviced) can flow through the base-emitter junction and possibly burn out the transistor.

Low-gain problems. As we have noted, low gain in a feedback amplifier can also result in distortion. That is, if gain is normal in a feedback amplifier, some distortion can be overcome. With low gain, the feedback may not be able to bring the distortion within limits. Of course, low gain by itself is sufficient cause to troubleshoot an amplifier (with or without feedback).

Assume, for example, the classic failure pattern of a solid-state feedback amplifier that was working properly but that now has a decrease in output of about 10 percent. This indicates a general deterioration of performance rather than a major breakdown.

Keep in mind that most feedback amplifiers have a very high open-loop gain that is set to some specific value by the ratio of resistor values (feedback-resistor value to load-resistor value). If the closed-loop gain is low, it usually means that the open-loop gain has fallen far enough so that the resistors no longer determine the gain. For example, if the ac beta of Q_2 shown in Fig. 3-25 is lowered, the open-loop gain is lowered. Also, the lower beta lowers the input impedance of Q_2, which, in turn, reduces the effective value of the load resistor for Q_1. This also has the effect of lower overall gain.

In troubleshooting such a situation, if wave forms indicate low gain and element voltages are normal, you should try replacing the transistors. Of course, you must never overlook the possibility of open or badly leaking emitter-bypass capacitors. If the capacitors are open or leaking (acting as a resistance in parallel with the emitter resistor), there will be considerable negative feedback and little ac gain. Of course, a completely shorted emitter-bypass capacitor produces an abnormal dc voltage indication at the transistor emitter.

Distortion problems. As we have discussed, distortion can be caused by improper bias, overdriving (too much gain), or underdriving (too little gain, preventing the feedback signal from countering the distortion). One problem often overlooked in a feedback amplifier with a pattern of distortion trouble is overdriving resulting from transistor leakage. (The problem of transistor leakage is discussed further in Sec. 3-2.16.)

Generally, it is assumed that the collector-base leakage will reduce gain because the leakage is in opposition to the signal-current flow. Although this is true in the case of a single stage, it may not be true when more than one feedback stage is involved.

Whenever there is collector-base leakage, the base assumes a voltage nearer to that of the collector (nearer than is the case without leakage). This increases both transistor forward bias and transistor-current flow. An increase in the transistor current causes a lower h_{ib} (ac input resistance, grounded-base configuration), which causes a reduction in common-emitter input resistance, which may or may not cause a gain reduction (depending on where the transistor is located in the amplifier).

If the feedback amplifier *is direct-coupled*, the effects of feedback are increased. This is because the operating point (base bias) of the following stage is changed, possibly resulting in distortion. For example, in Fig. 3-25, the collector of Q_1 is connected directly to the base of Q_2. If Q_1 starts to leak (or if the collector-base leakage increases with age), the base of Q_2 (as well as the collector of Q_1) will shift its Q point (no-signal voltage level at the base).

3-2.16 Effects of Transistor Leakage on Amplifier Gain

When there is considerable leakage in a solid-state amplifier, the gain is reduced to zero, and/or the signal wave form is drastically distorted. Such a condition also produces abnormal wave forms and transistor voltages. These indications make troubleshooting easy or at least relatively easy. The troubleshooting problem becomes really difficult when there is just enough leakage to reduce amplifier gain but not enough leakage to distort the wave form seriously or produce transistor voltages that are way off.

Collector-base leakage is the most common form of transistor leakage and produces a classic condition of low gain (in a single stage). When there is any collector-base leakage, the transistor is forward-biased, or the forward bias is increased, as shown in Fig. 3-28.

Collector-base leakage has the same effect as a resistance between the collector and the base. The base assumes the same polarity as the collector (although at a lower value), and the transistor is forward-biased. If leakage is sufficient, the forward bias can be enough to drive the transistor into or near saturation. When a transistor is operated at or near the saturation point, the gain is reduced (for a single stage), as shown in Fig. 3-29.

If the normal transistor-element voltages are known (from the service literature or from previous readings taken when the amplifier was operating properly), excessive transistor leakage can be spotted easily because all the transistor voltages will be off. For example, in Fig. 3-28, the base and emitter will be high, and the collector will be low (when measured in reference to ground).

If the normal operating voltages are not known, the transistor can appear to be good because all the voltage *relationships* are normal. That is, the collector-base junction is reverse-biased (collector more positive than base for an *npn*), and the emitter-base junction is forward-biased (emitter less positive than base for an *npn*).

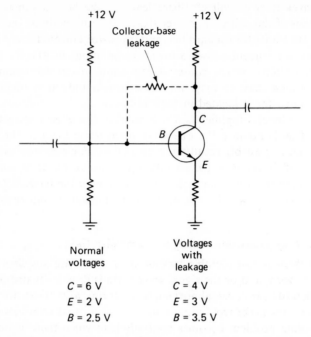

Normal voltages	Voltages with leakage
C = 6 V	C = 4 V
E = 2 V	E = 3 V
B = 2.5 V	B = 3.5 V

FIGURE 3-28 Effect of collector-base leakage on transistor-element voltages

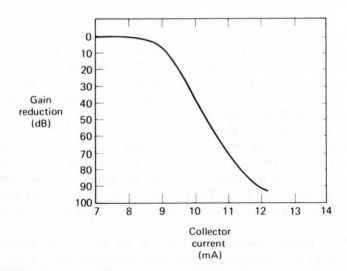

FIGURE 3-29 Relative gain of solid-state amplifier at various average collector-current levels

A simple way to check transistor leakage is shown in Fig. 3-30. Measure the collector voltage to ground. Then, short the base to the emitter, and remeasure the collector voltage. If the transistor is not leaking, the base-emitter short will turn the transistor off, and the collector voltage will rise to the same value as the supply. If there is any leakage, a current path will remain (through the emitter resistor, emitter-base short, collector-base leakage path, and collector resistor). There will be some voltage drop across the collector resistor, and the collector will have a voltage at some value lower than the supply.

Note that most meters draw current, and this current passes through the collector resistor. This can lead to some confusion, particularly if the meter draws heavy current (has a low ohms-per-volt rating). To eliminate any doubt, connect the meter to the supply through a resistor with the *same* value as the collector resistor. The drop, if any, should be the same as it is when the transistor is measured to ground. If the drop is much different (lower) when the collector is measured, the transistor is leaking.

For example, assume that in the circuit shown in Fig. 3-30 the supply is 12 V, the collector resistance is 2 kΩ, and the collector measures 4 V with respect to ground. This means that there is an 8 V drop across the collector resistor and a collector current of 4 mA (8/2,000 = 4 mA). Normally, the collector is operated at about one-half the supply voltage (in this case, 6 V).

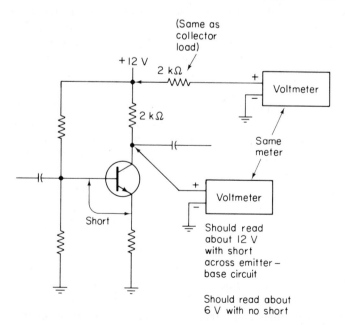

FIGURE 3-30   Checking for transistor leakage in amplifier circuit

However, simply because the collector is at 4 V instead of 6 V does not make the circuit faulty. Some circuits are designed that way.

In any event, the transistor should be checked for leakage with the emitter-base short test shown in Fig. 3-30. Now, assume that the collector voltage rises to 10.5 V when the base and emitter are shorted. This indicates that the transistor is cutting off but that there is still some current flow through the collector resistor, about 1 mA ($2/2,000 = 1$ mA).

A current flow of 1 mA is high for a meter. However, to confirm a leaking transistor, connect the same meter through a 2 kΩ resistor (same as the collector load) to the 12 V supply (preferably at the same point where the collector resistor connects to the power supply). Now, assume that the indication is 11.7 V through the external resistor. This indicates that there is some transistor leakage.

The amount of transistor leakage can be estimated as follows: $11.7 - 10.5 = 1.2$ V drop and $1.2/2,000 = 0.6$ mA. However, from a practical troubleshooting standpoint, the presence of any current flow with the transistor supposedly cut off is sufficient cause to replace the transistor.

3-2.17 Example of Audio-Amplifier Troubleshooting

This step-by-step troubleshooting problem involves locating the defective component in a solid-state audio amplifier and then repairing the trouble. Because this amplifier is designed to perform a single electronic subfunction (AF amplification), the amplifier can be considered a single-unit multicircuit piece of equipment. Therefore, you can skip the troubleshooting step of localizing the trouble to a functional unit (discussed in Chapter 1).

General instructions. The schematic diagram for the audio amplifier is shown in Fig. 3-31. There is no servicing block diagram. This is typical for this sort of simple equipment. No test points (as such) are given on the schematic diagram, and the voltage information is incomplete. There is no resistance information. The only service literature supplied with equipment is the schematic, and a note (on the schematic) stating that the output is 25 W across a 4 Ω load (loudspeaker) when a 0.1 V (100 mV) signal is introduced at the input (across R_1) and R_1 is at midrange.

Using this fragmentary information (which is probably more than you will get in practical troubleshooting situations) you can pencil in the logical test points, as shown in Fig. 3-31. The test points are logical because they show the input and output of each stage. Note that the test points show *linear, separating,* and *meeting* signal paths (discussed in Chapter 1).

Also, using what you know, the input to be introduced at test A should be 0.1 V at some audio frequency. Thus, you can connect an audio generator to test point A (as shown in Fig. 3-32) and set the generator to produce 0.1 V (100 mV) at a frequency of 1,000 Hz (or some other frequency in the audio range). Under these conditions, the output voltage at test point H should

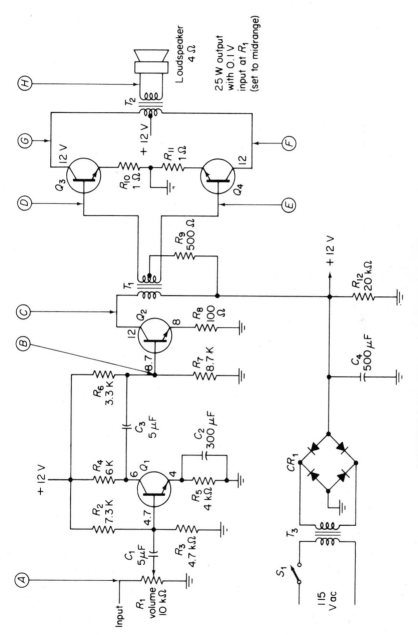

FIGURE 3-31 Schematic diagram of simple audio amplifier

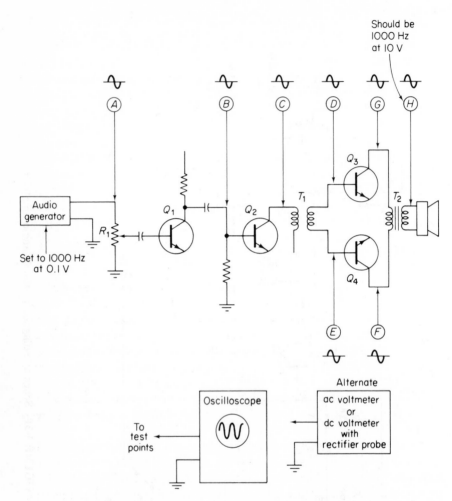

FIGURE 3-32   Basic test connections for troubleshooting simple audio amplifier

be about 10 V. This voltage is found because the output is supposed to be 25 W across a 4 Ω load and $E = \sqrt{PR}$, or $E = \sqrt{25 \times 4} = \sqrt{100} = 10$ V.

You have no idea what the signals at other test points are. But you do know that (with an appropriate signal introduced at A) the remaining test-point signals should be sine waves at the same frequency. There will probably be considerable voltage gain at points B and C, but you can only guess how much.

You can monitor each of the test points with an ac voltmeter, a dc voltmeter with a rectifier probe, or a scope. We shall use the scope because it will show any really abnormal distortion at each of the test points, as well as any gain.

Armed with this information, you are now ready to begin the troubleshooting effort. You may make notes as you go along if that will help. However, keep in mind that each troubleshooting problem is always slightly different from the last. There is no surefire step-by-step procedure that will fit every situation.

Determine the symptoms. A trouble of no audio output from the speaker is reported to you, and you then check the equipment yourself (never trust anyone). You find that the symptom of no output is correct when the volume control R_1 is set at its midrange position (which is the normal operating position for R_1). However, by rotating R_1 fully clockwise (for maximum volume), you note that there is a very weak tone from the speaker. Thus, the no-audio-output symptom is not correct, but it is obvious that the amplifier is not operating properly. (With R_1 at nearly full ON and a 0.1 V input, the output should be over 25 W, and the tone would probably burst your eardrums.)

Reset R_1 to its normal midrange position, and make your first decision.

You could decide that the audio generator is bad and that there is no input signal at test point A. This is not logical. When the volume control is turned to maximum volume, there is a weak tone from the speaker. Thus, although this output is weak, it does tell you that there is an input signal present. If there were no tone whatsoever, then you could say that the audio generator *possibly* could be bad. Just to satisify yourself, connect the scope across the input terminals (test point *A*), and observe the input wave form. It should be a sine wave at a frequency of 1,000 Hz and with an amplitude of about 0.1 V.

You could decide that the power supply is defective. This is slightly more logical than the decision concerning the generator. If the power supply is bad, you will get a symptom of no output tone whatsoever when R_1 is set to maximum volume. Because there is a weak output, *possibly* resulting from stray coupling around the defective stage or from a weak transistor, the power supply is working. Of course, the power supply could be producing a low voltage to one or more stages. To confirm this, you must check the voltage at each of the stages. However, at this point in the troubleshooting, remember that you are trying to determine the symptoms, not isolate the trouble.

You should decide to check the output of the circuit group. This is the first step in isolating the trouble to a circuit.

Isolate the trouble to a circuit. This is accomplished by checking the wave form at the output (test point *H*) of the circuit group. The test connections are shown in Fig. 3-33. If the amplifier is operating properly, there should be a sine wave at *H* with an amplitude of about 10 V when R_1 is set

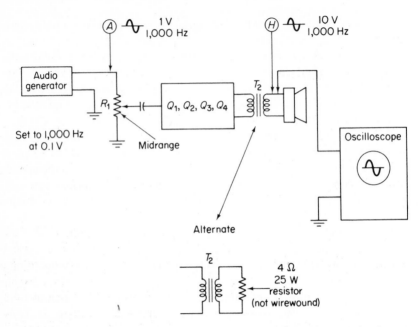

FIGURE 3-33   Test connections used to isolate trouble to circuit group (audio amplifier)

to midrange. If the sine wave is present but there is no tone in the speaker, then the speaker is suspect. If a replacement speaker is available, check by substitution. If no speaker is available, connect a 4 Ω resistor (as shown in Fig. 3-33) across the secondary terminals of T_2 (to substitute as the load), and observe the sine wave at H. If the sine wave is correct with a substitute load, the speaker is at fault. *Do not operate the amplifier without a load* (either the speaker or the resistor). To do so could damage the transistors.

If the sine wave is absent at H with R_1 set to midrange, you can place a bad-output bracket at test point H (as discussed in Chapter 1). We have a good-input bracket at A and a bad-output bracket at H; now it is time to make another decision.

You could decide that the output transformer T_2 is defective. This is possible but not logical. First of all, the trouble must be isolated to a single circuit by checking the input and output points. Thus far, you know there is a normal input at test point A (the input to the circuit group) and an abnormal output at test point H (the output of the circuit group). The trouble is located somewhere between these points, somewhere within the four circuits preceding test point H, *possibly* even in T_2. If the trouble was in T_2, and you found it immediately, it would be a lucky guess, not logical troubleshooting. Further testing must be done before you can say definitely that any one circuit is defective.

You could decide to use the half-split technique and make the next check at test points D or E. Assume that you monitored point D and found no signal present (with R_1 at midrange). This is definitely an abnormal condition, but it proves very little. The problem could be associated with the circuits of Q_1, Q_2, or Q_3. Even the Q_4 circuit is not definitely eliminated (until you check at E). The fact that by chance you have selected a test point yielding an abnormal signal does not make your procedure correct.

You could decide to use the half-split technique and make the next check at test point B. This is a fairly logical choice. You must make a test at B sometime during the troubleshooting sequence. If the signal at B is abnormal, you have isolated the trouble to the Q_1 circuit (in one lucky jump). However, if the signal is normal at B, you are left with three possibly defective circuits (Q_2, Q_3, or Q_4). There is a more logical choice.

You should decide to use the half-split technique and make the next check at test point C. This is the most logical choice because you have isolated the trouble to one-half of the equipment (or two circuits) in one jump. If the signal at C is abnormal, the trouble is in Q_1 or Q_2. If the signal at C is normal, the trouble is isolated to Q_3 or Q_4. (Note that the primary winding of T_1 is considered part of the Q_2 circuit but that the secondary winding of T_1 is part of the Q_3–Q_4 circuit.)

Keep in mind that the terms *normal* and *abnormal* applied to the signals at test points B to G are arbitrary and relative. The service literature does not tell you the signal amplitudes or the correct wave forms. However, it is reasonable to assume that all the signals are sine waves (at least ac voltages at the frequency of the signal introduced in test point A). It is also reasonable to assume that there is some voltage gain at each test point as you proceed along the signal path. With an input of 0.1 V, the service literature says you can expect an output of 10 V. This is a voltage gain of 100. Most of the gain (at least half, probably more) is obtained in the Q_1 and Q_2 circuits (test points B and C) because the Q_3 and Q_4 circuits are essential power amplifiers. Similarly, the circuit of Q_1 should show more voltage gain than the circuit of Q_2, because the emitter resistor of Q_1 is bypassed by C_2. Thus, the *difference in gain* should be greater between points A and B than between points B and C. However, because exact values are not available, you must ultimately isolate the trouble with voltage-resistance checks, component checks, and the like.

Now, assume that you make the check at test point C and find the signal abnormal. If you are paying any attention, your next logical step is to monitor the signal at test point B. There is no reason to monitor any other test point under these conditions. The signals in the paths beyond point C will be abnormal if point C is abnormal.

Now, assume that you make the check at test point B and find the signal normal. You have now isolated the trouble to a circuit (the Q_2 circuit), and you have done so in three logical jumps (from test point H to C to B). Your next step is to locate the specific trouble in the Q_2 circuit.

Locate the specific trouble. As discussed in Chapter 1, the first step in locating the specific trouble after the circuit has been isolated is to perform an inspection using the senses. As well as can be determined, the transistor is good because there is no evidence of physical damage (look to see if this is true) and there is no evidence of overheating (touch the transistor; it should *not* be hot). There is no indication of burning components, (no characteristic burning smell), and there are no obvious physical defects. You can conclude, therefore, that the inspection using the senses points to no outward sign of where the trouble is located. You must then rely on test procedures to locate the defective component.

Figure 3-34 shows both the physical relationship of the Q_2 circuit parts and the point-to-point wiring. This illustration is similar to that provided in well-prepared service literature. Count yourself lucky if you have such data when troubleshooting all electronic equipment.

It is now time to make another decision concerning your next step in troubleshooting.

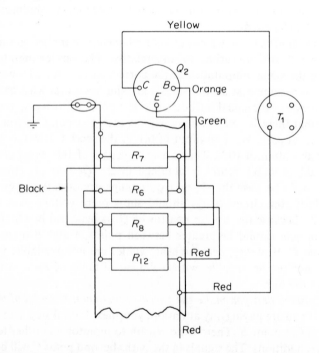

FIGURE 3-34   Physical relationship of Q_2 circuit parts and point-to-point wiring (practical wiring diagram)

You could observe the wave form at the emitter of Q_2. This would be of little value. With a normal amplifier circuit, the emitter wave form is similar to the collector wave form (test point *C*) except that the emitter usually shows lower amplitude. Because there is a low-amplitude wave form (or no amplitude) at the collector, you will find nothing of value at the emitter.

You could decide that transformer T_1 *is defective.* To prove your assumption, you could measure the voltage at the collector of Q_2. The schematic (Fig. 3-31) shows that the voltage should be 12 V. If there is no voltage at the collector of Q_2 or the voltage is very low, T_1 is probably defective (the primary is probably open or shorted). Your assumption is logical, but you are ahead of yourself. What if the voltage at the collector of Q_2 is correct? Then, T_1 is probably good (at least the primary is not open or shorted).

The sequence you have just performed should have taught you one thing: Do not make hasty decisions with regard to faulty components. Make your decisions after you have gained enough information from conducting the proper tests and measurements. There is a more logical choice than faulting transformer T_1 immediately.

You should check the voltages at each element of Q_2. Figure 3-35 shows the test connections for measuring the voltages. Note that the negative (−) terminal of the voltmeter is connected to a ground terminal on the chassis and that the positive (+) terminal is connected to each of the transistor elements in turn. The reason for this is that the power supply is positive with respect to ground, as indicated on the schematic.

Figure 3-31 shows that the voltages should be +8, +8.7, and +12 V,

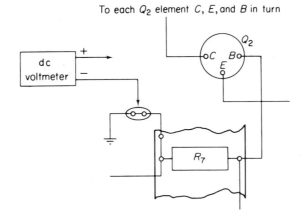

FIGURE 3-35   Test connections for measuring voltages at Q_2 elements

respectively, for the emitter, base, and collector. If *all* the voltages are normal or nearly so, it is logical to assume that transistor Q_2 is at fault. You can make an in-circuit test of the transistor as described in Sec. 3-1.4, or you can substitute a known-good transistor, whichever is most practical. (If the transistor fails the in-circuit test, you must try substitution.)

If the voltage at one or more of the Q_2 elements is abnormal, you must now make resistance checks. As you know from the discussion in Chapter 1, this is done by measuring the resistance from each element of the transistor to ground, using the resistance charts supplied in the service literature as a reference. But in this case, you have no resistance charts. Furthermore, the schematic does not give enough information for you to calculate the resistance-to-ground for each Q_2 element. A possible exception is the emitter. Here, the resistance should be 100 Ω because only R_8 (a 100 Ω resistor) is connected to the emitter. Both the base and the collector have several resistances in parallel. In any event, the correct resistance-to-ground will be only a wild guess.

Under these circumstances, continuity and resistance checks are your best bet. Let us examine each element of Q_2 in turn.

The collector of Q_2 is connected to the power supply through the primary winding of T_1. To check continuity in this line, disconnect the collector lead, and make the connections shown in Fig. 3-36. Remember that you do not know the resistance of the T_1 primary winding, but you should have a continuity indication, *probably* in the order of a few ohms. If the ohmmeter shows an infinite resistance (with the range selector set on one of the high-

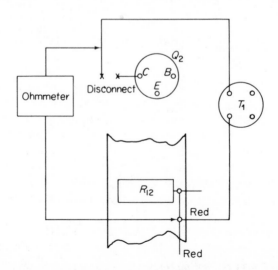

FIGURE 3-36   Test connections for checking continuity of Q_2 collector circuit

resistance scales), the T_1 primary winding is probably open. If the ohmmeter resistance is zero (on the lowest scale), the T_1 primary winding is probably shorted. Keep in mind that you can skip this resistance measurement if the collector of Q_2 shows a normal voltage (about $+12$ V).

The emitter of Q_2 is connected to ground through R_8. To check continuity here, disconnect the emitter lead, and make the connections shown in Fig. 3-37. The resistance should be about 100 Ω. Keep in mind that because all resistors have some tolerance, the reading will probably never be exactly 100 Ω. Again, a high or infinite reading indicates an open; whereas a low or zero reading indicates a short. From a practical standpoint, resistors usually do not short, but they do open.

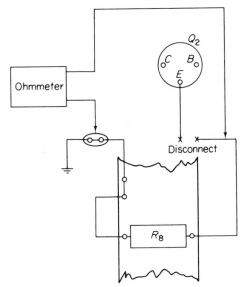

FIGURE 3-37 Test connections for checking continuity of Q_2 emitter circuit

The base of Q_2 is connected to the power supply through R_6. To check continuity in this line, disconnect the lead, and make the test connections shown in Fig. 3-38. This will remove any parallel resistance from R_7 or Q_2. The resistance should be 3.3 kΩ.

The base of Q_2 is connected to ground through R_7. To check continuity in this line, disconnect the lead, and make the test connections shown in Fig. 3-39. This will remove any parallel resistance from R_6 or Q_2. The resistance should be 8.7 kΩ.

Now, assume that the continuity-resistance checks show that R_8 is open. As discussed in Chapter 1, your next step is to make the necessary repairs and perform an operational check.

Repairs and operational check. After you have reviewed all the data and are satisfied that you have located the specific cause of the trouble, you

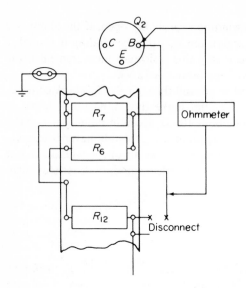

FIGURE 3-38 Test connections for checking continuity of Q_2 base circuit to power supply

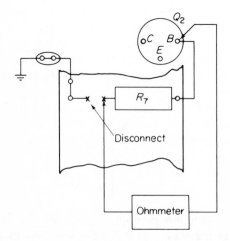

FIGURE 3-39 Test connections for checking continuity of Q_2 base circuit to ground

should then repair the trouble. In this case, you should replace emitter resistor R_8 with a known-good resistor. What is your next step?

You could turn on the power and make wave-form and voltage measurements at Q_2. Your first step after repairing the trouble should not be to turn on the power to the equipment. Rather, you should first verify that the repair you have made is good. That is, because this trouble was an open emitter circuit, you should check out the emitter circuit (continuity measurement) to be sure that the repair you have made is complete.

You should, before turning on the power, measure the resistance of R_8 and check continuity from the emitter of Q_2 to ground. Because this trouble was an open emitter circuit, you should first check out the emitter circuit for proper resistance and continuity. If by some strange chance the emitter resistance is still abnormal, you can assume that the repair you have made is not proper (or that you are incorrectly interpreting the resistance reading). Whatever the reason, you must take another look at your procedure. However, if the resistance and continuity from the emitter to ground is normal, you can be reasonably certain that you have properly repaired the trouble.

Once proper resistance and continuity are established, you can turn on the power and make an operational check. What is the first thing you would do while making this check?

You could make voltage measurements at all elements of Q_2. It is unnecessary to make these measurements at this time. You are now performing the operational check, not trying to isolate trouble to a defective branch of a circuit. You are fairly sure that the trouble is repaired; now, you simply want to verify this fact.

You should check to be sure that all controls and switches, including those of the test equipment, are set for normal operation. Always make sure that all switches and controls are first set for *normal operation* when you are performing an operational check. If the volume control (R_1 in Fig. 3-31) is set for minimum volume, there will be no sound from the speaker and no wave forms available at the test points, even though the amplifier is operating properly. Similarly, if the output control on the audio generator (the signal-injection source) is accidently set for minimum during troubleshooting, there will be no sound and no wave forms. Because these wrong control settings can be misinterpreted as equipment trouble, it is very important to set *all* controls for normal operation before attempting to perform the operational check.

3-3. RECEIVER TROUBLESHOOTING

Before presenting a step-by-step example of receiver troubleshooting, we shall discuss basic test and alignment procedures for receivers (both AM and FM).

3-3.1 Testing Receiver-Tuning Circuits

It is possible to test and adjust receiver-tuning circuits using a meter and signal generator. Both AM and FM receivers require alignment of the IF and RF amplifiers. An FM receiver also requires alignment of the detector stage (discriminator or ratio detector). If a complete receiver is being tested and the receiver includes an AVC circuit, the AVC must be disabled (by means of a fixed bias, typically on the order of a few volts).

FM detector alignment. The circuits for FM detector alignment are shown in Fig. 3-40. Note that both ratio-detector and discriminator circuits are given. In both cases, the signal to be injected is an unmodulated RF signal tuned to the intermediate frequency of the receiver. The signal can be injected at the input of the first IF stage or at the primary winding of the detector transformer (which is also the final or output transformer of the IF stages).

With either circuit, a dc voltmeter is required to monitor the detector output. However, the points at which the meter is connected into the circuit are different for each type of detector, as shown in Fig. 3-40.

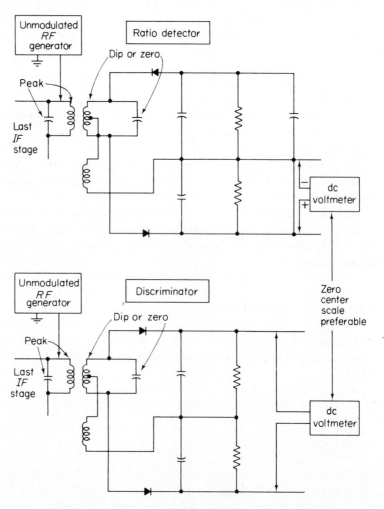

FIGURE 3-40   Connections for FM detector alignment

If necessary, disable the AVC line (which may also be called an *AGC line*). This is done by applying a fixed dc voltage between the line and ground, as shown in Fig. 3-41. In the absence of specific values, always use the same polarity as the normal AVC voltage and a value that is higher than the average AVC voltage (about twice the average value). For example, if the AVC line usually varies between 0 and −1 V, use a value of −2 V.

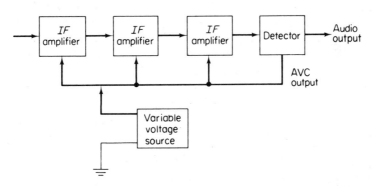

FIGURE 3-41 Connections for disabling AVC line

Adjust the signal-generator frequency to the intermediate frequency (typically 10.7 MHz for household FM receivers). Use an *unmodulated* output from the signal generator.

Adjust the *secondary winding* of the transformer (either the capacitor or a tuning slug within the winding) for a *zero reading* (or a dip) on the meter. Adjust the transformer slightly each way, and make sure that the meter moves smoothly above and below the exact zero mark (or the minimum dip point). (A meter with a zero-center scale, as described in Sec. 2-2.2, is most helpful when adjusting FM detectors).

Adjust the signal generator to some point below the intermediate frequency (to 10.625 MHz for an FM detector with a 10.7 MHz IF). Note the meter reading.

Adjust the signal generator to some point above the intermediate frequency *exactly equal* to the amount set below the intermediate frequency. For example, if the generator is set to 0.075 below the intermediate frequency, set the generator to 0.075 above the intermediate frequency (or to 10.625 and 10.775 MHz, respectively).

The meter reading should be *approximately the same* on both sides of the intermediate frequency except that the polarity is reversed. For example, if the meter reading is 7 scale divisions below zero and 7 scale divisions above zero (on a zero-center meter), the FM detector is balanced. If an FM detector cannot be balanced, the fault is usually a *serious mismatch* in diodes or other components.

Return the generator to the intermediate frequency (10.7 MHz), and adjust the *primary winding* of the transformer for *maximum* or *peak* reading on the meter. This sets the primary winding at the correct resonant frequency of the IF stages.

AM and FM alignment. The alignment procedures for the IF amplifier stages of an AM receiver are essentially the same as those used for an FM receiver. However, the meter must be connected at different points in the corresponding detector, as shown in Fig. 3-42. In either case, the meter is set to measure direct current, and the RF probe is not used.

In those cases in which the IF stages are being tested without a detector (e.g., during design or after extensive troubleshooting), an RF probe is required. As shown in Fig. 3-42, the RF probe is connected to the secondary winding of the final IF output transformer.

If necessary, disable the AVC line as described for FM detector alignment. Then, set the meter to measure direct current, and connect it to the appropriate test point (with or without an RF probe, as applicable). Adjust the generator frequency to the receiver's intermediate frequency (typically 10.7 MHz for FM and 455 kHz for AM household receivers). Use an *unmodulated* RF signal.

Adjust the windings of the IF transformers (capacitor or tuning slug) in turn, starting with the *last* stage and working toward the first stage. Adjust each winding for *maximum* reading. Repeat the procedure to make sure that there is no interaction between adjustments (usually there is some interaction).

AM and FM RF amplifier alignment. The alignment procedures for the RF stages (RF amplifier, local oscillator, mixer-converter) of an AM receiver are essentially the same as those used for an FM receiver. Again, it is a matter of connecting the meter to the appropriate test point. The same test points used for IF alignment can be used for aligning the RF stages, as shown in Fig. 3-43. However, if an individual RF stage is to be aligned, the meter must be connected to the secondary winding of the RF stage output transformer through an RF probe.

If necessary, disable the AVC line as described for FM detector alignment. Set the meter to measure direct current, and connect it to the appropriate test point (with or without an RF probe, as applicable).

Adjust the generator frequency to some point near the high end of the receiver's operating frequency (typically 107 MHz for an FM broadcast receiver and 1,400 kHz for an AM broadcast receiver). Use an unmodulated output from the signal generator.

Adjust the trimmer of the RF stage for *maximum* reading on the meter.

Adjust the generator frequency to the low end of the receiver's operating frequency (typically 90 MHz for FM and 600 kHz for AM).

Adjust the trimmer of the oscillator stage for *maximum* reading on the meter.

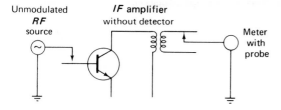

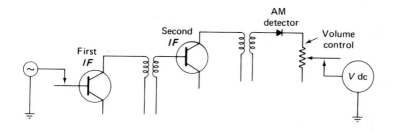

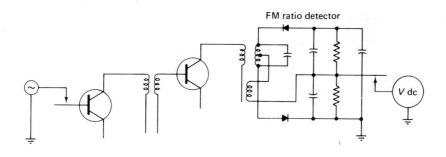

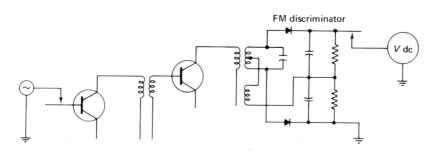

FIGURE 3-42 Connections for IF alignment of AM and FM receivers

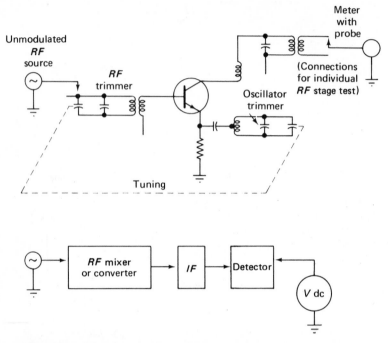

FIGURE 3-43   Connections for FM amplifier and local-oscillator alignment of AM and FM receivers

Repeat the procedure to make sure that the resonant circuit tracks across the receiver's entire tuning range.

3-3.2 Example of Receiver Troubleshooting

This step-by-step troubleshooting problem involves locating the defective component in a solid-state radio receiver and then repairing the trouble. Because this receiver is considered a single-unit multicircuit piece of equipment, you can skip the *localize* step of troubleshooting.

General instructions. The schematic diagram for the receiver is shown in Fig. 3-44. Note that the audio portion of the receiver is identical to that described for the audio-amplifier troubleshooting example (Sec. 3-2.17). Also note that an IC is used for the IF stages of the receiver. Thus, if a fault is traced to any circuit within the IC, it is necessary to replace the entire IC unit. All the IC terminals are identified on the schematic, but none of the terminals are shown for the other components.

As discussed, servicing block diagrams are not always available for commercial equipment. In fact, because of its almost universal usage, the common household radio often comes without a schematic diagram. Even when a schematic is available, the test points are seldom shown.

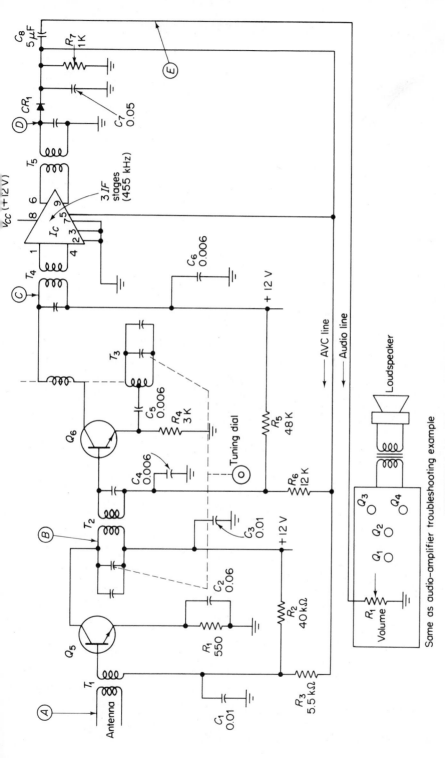

FIGURE 3-44 Schematic diagram of AM receiver

Same as audio-amplifier troubleshooting example

When test points are not shown on the schematic, good practice would insist that you determine where the test points are located before troubleshooting the equipment. In general, use the collector and bases of the transistors (or plates and grids of vacuum tubes) when test points are not indicated on the schematic. It is entirely possible to troubleshoot the receiver (or any similar equipment) using only this minimum information. Consider the following:

The user identifies the receiver as a broadcast radio, and dial markings on the front panel show that the frequency range is from 550 to 1,600 kHz. The schematic shows that the IF frequency is 455 kHz (which is standard for AM broadcast radios). From a troubleshooting standpoint, this establishes certain conditions.

You know that an RF signal can be introduced at test point A. The signal must be in the range of 550 to 1,600 kHz. The same is true for test point B. However, if you inject a signal at B, it must be much greater in amplitude because Q_5 acts as an RF amplifier and normally supplies considerable voltage gain. If you are monitoring the signal at B, it should be identical to that at A except for amplitude.

Signals at test points C and D will be at 455 kHz. From a monitoring standpoint, the signal at D should be much larger in amplitude than that at C because of the gain produced in the IC amplifier stages. The signal at test point C may also include the local-oscillator frequency developed by transformer T_3. You do not know this frequency, but you can assume that it runs from 1,005 to 2,055 kHz because the local oscillator *usually* operates at a frequency equal to the incoming RF signal plus the intermediate frequency (or $550 + 455 = 1,005$).

Transformer T_1 is untuned. Transformer T_2 is tuned to the incoming RF frequency by the tuning dial. Transformer T_3 is also tuned by the dial but is at a frequency 455 kHz higher than the incoming frequency. Transformers T_4 and T_5 are both tuned to 455 kHz.

No voltage or resistance information is available except for the $+12$ V supply voltage indicated on the schematic. The collectors of Q_5 and Q_6 are thus at (or near) $+12$ V, as is the supply IC package (terminal 8). All remaining voltages are a mystery.

The signal at test point E is in the audio range. Thus, from the standpoint of signal injection, you must inject an AF signal at E. If test point E is used for signal tracing, you must use an ac meter or a dc meter with a probe (or a scope) to monitor the signal at E. Also, if the incoming RF signal (either the broadcast signal or from an RF generator connected to some point ahead of E) is unmodulated, there will be no signal at E, even though the receiver is operating properly. Thus, if an RF generator is connected to A and you expect to monitor the signal at E with a scope or an ac meter, you must modulate the RF signal-generator output.

With this limited information scrounged from the schematic (and a basic knowledge of receivers), you are ready to start the troubleshooting.

Determine the symptoms. The user reports that no stations can be heard on any frequency. You confirm this symptom, but you notice that noise can be heard at the speaker when the volume control (in the audio-amplifier section) is set for maximum volume. What is the most logical test point at which to begin troubleshooting?

You could check the supply voltage $(+12\ V)$ *at various points throughout the circuit.* This idea has some merit. It is *possible* that a low supply voltage could reduce signals to the point where they cannot be heard on the speaker. However, you are going at the problem in a hit-or-miss fashion. Consider the following examples:

Assume that the trouble is an open primary winding in T_2. This will show up as no voltage at the collector of Q_5 and will trace the trouble quickly (if you were lucky enough to check the collector of Q_5 first).

Now, assume that the trouble is a shorted primary winding in T_2. You will get a voltage indication at the collector of Q_5. But you do not know what voltage is correct; you know only that it should be about $+12$ V. If you find a voltage near $+12$ V, you must assume that T_2 is good, and you will miss the trouble.

You could monitor the signal at test point D. This is a more logical choice because you have effectively split the equipment in half. However, to monitor the signal at D, you must use a scope or ac voltmeter (or dc meter with probe). Also, you must introduce a signal (preferably a modulated RF signal) at some point ahead of D (probably at A) or rely on a broadcast signal, and you must tune the receiver to the appropriate frequency (of the generator or broadcast signal).

You should inject an audio signal at test point E. This is the most logical place for the first test. You are simultaneously eliminating as many functions as possible by dividing the receiver into an RF function and an audio function. The results of this test should therefore isolate the trouble to either the audio or the RF circuits. Note that this is a convenient test point for any receiver.

With an audio signal applied at E, you should hear a tone from the speaker. If you do not hear a tone, the audio section is defective, and you should begin the troubleshooting as described in Sec. 3-2.17.

Now, assume that a signal is easily heard from the speaker, indicating that the audio circuits are performing satisfactorily. You should have suspected this when you noticed that the volume control affected the amplitude of the noise.

Now that the audio circuits are cleared of trouble, you must select another test point and make tests to isolate the trouble further. Because test point A is for signal injection and you have eliminated everything beyond test point E, you only have three test points left: B, C, and D. Which do you choose?

You could choose test point B. If you inject a signal at B and the response is good (i.e., if you inject a modulated RF signal and hear the tone on the speaker), you have isolated the trouble to the RF voltage amplifier Q_5, and you are also very lucky. If the response is bad with a signal injected at B, you still have many circuits to check. You could also monitor the signal at B, but this requires two test instruments (unless you rely on broadcast signals).

You could choose test point D. If you inject a signal at D and the response is bad (i.e., if you inject a modulated RF signal and do not hear the tone on the speaker), you have isolated the trouble to the detector CR_1 circuit, and you are again very lucky. If the response is good with a signal injected at D, you still have many circuits to check. Again, you could monitor the signal at D, but this requires two test instruments.

There is another problem with test point D. You could choose to inject an *audio* signal at D. If CR_1 and the associated circuit parts are good, the audio signal should pass as if it were injected at E. However, it is possible that a shorted or badly leaking CR_1 will also pass the straight audio signal but will not pass the audio portion of the modulated RF signal.

You should choose test point C. By choosing C, you have effectively split the remaining circuits in half. (Test point C is halfway between the antenna and test point E). You are on the right track. But do you inject a signal at C, or monitor the signal at C? Many technicians will give you an argument either way. But because injection is simpler (in this case), you choose to inject a 455 kHz modulated signal at C, and you find that there is a good response (i.e., you hear a tone on the loudspeaker, and there is plenty of volume).

The trouble is now isolated to two circuits, the RF voltage amplifier Q_5 and the converter Q_6. To isolate the trouble further, the obvious next step is to inject a signal at B. (Use a modulated RF signal at the frequency to which the receiver is tuned.) If the response is good, the trouble is isolated to the RF voltage amplifier. However, to make the problem more difficult, assume that the response is bad with a signal injected at B. The trouble is now isolated to the converter.

Because you lack comprehensive service information (not an unusual situation), you must troubleshoot the converter by using some logical guesswork (unless you simply want to replace every part in the converter circuit).

You can start by checking the voltage at each terminal of Q_6. You do not know the exact voltages, but you can make some close guesses. The collector should be near $+12$ V, because there is little voltage drop in the windings of T_3 and T_4. The value of R_6 is about one-fifth that of R_5 plus R_6; thus, the voltage drop across R_6 is about one-fifth of the full 12 V across both resistors (ignoring the small drop across R_7). The junction of R_5 and R_6 (to which the base of Q_6 is connected) is thus at about 2 or 3 V (approximately one-fifth of 12 V). Because Q_6 is an *npn* and should normally be forward-biased, the emitter should be less positive than the base by about 0.5 to 1 V. Thus, if the base is at $+3$ V, the emitter will be at about $+2.5$ V.

Keep in mind that Q_6 is an oscillator and that oscillator-bias relationships require special consideration. Oscillator-bias problems are discussed in Sec. 3-4.

Now, assume that all the voltages are good, or that they at least appear logical. You can then make an in-circuit test of the transistor as described in Sec. 3-1.4, or you can substitute the transistor, whichever is most convenient.

Because Q_6 is an oscillator, you have one more troubleshooting check available before you make point-to-point checks of each component. If Q_6 is operating properly, it should be oscillating and producing an RF signal. You can monitor this signal at several points in the circuit. However, a convenient point is at the windings of T_3, as shown in Fig. 3-45. Use a scope or meter with an RF probe.

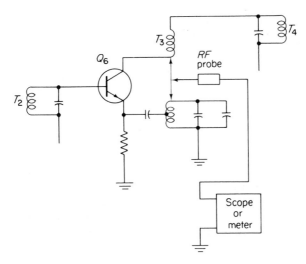

FIGURE 3-45   Connections for checking Q_6 local-oscillator circuit operation

3-4. TRANSMITTER TROUBLESHOOTING

Before presenting a step-by-step example of transmitter troubleshooting, we shall discuss basic test and alignment procedures for transmitters. We shall also discuss problems that arise in troubleshooting oscillator circuits.

3-4.1   Oscillator Troubleshooting Problems

One of the problems in troubleshooting solid-state oscillator circuits is the bias arrangement. RF oscillators (the kind used in transmitters) are generally reverse-biased, so that they conduct on half-cycles. However, the transistor is initially forward-biased by dc voltages (through the bias-resistance network, as shown in Fig. 3-46). This turns the transistor on so that the collector circuit starts to conduct. Feedback occurs, and the transistor is driven into heavy conduction.

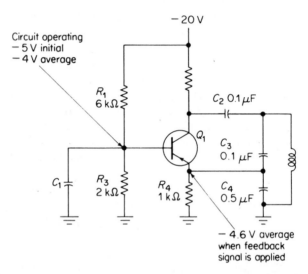

FIGURE 3-46   Class C RF oscillator (reverse-biased or zero-biased with circuit operating)

During the time of heavy conduction, a capacitor connected to the transistor base is charged in the forward-bias direction. When saturation is reached, there is no further feedback, and the capacitor discharges. This reverse-biases the transistor and maintains the reverse bias until the capacitor has discharged to a point where the fixed forward bias again causes conduction.

This condition presents a problem in the operation of class C solid-state RF oscillators. If the capacitor is too large, it may not discharge in time for the next half-cycle. In that case, the class C oscillator acts as a blocking oscillator, controlling the frequency by the capacitance and resistance of the circuit. If the capacitor is too small, the class C oscillator may not start at all. The same is true if the capacitor is leaking badly. From a practical troubleshooting standpoint, the measured condition of bias on a solid-state oscillator can provide a good clue to operation if you know how the oscillator is supposed to operate.

The oscillator of Fig. 3-46 is initially forward-biased through R_1 and R_3. As Q_1 starts to conduct and in-phase feedback is applied to the emitter (to sustain oscillation), capacitor C_1 starts to charge. When saturation is reached (or approached) and the feedback stops, capacitor C_1 then discharges in the opposite polarity, reverse-biasing Q_1. The value of C_1 is selected so that C_1 discharges to a voltage less than the fixed forward bias before the next half-cycle. Thus, transistor Q_1 conducts on slightly less than the full half-cycle. Typically, a class C RF oscillator such as the one shown in Fig. 3-46 conducts on about 140° of the 180° half-cycle.

While we are on the subject of bias, it should be noted that it is commonly assumed that transistor junctions (and diodes) start to conduct as soon as forward voltage is applied. This is *not* true. Figure 3-47 shows characteristic curves for three different types of transistor junctions. All three junctions are silicon, but the same condition exists for germanium junctions. None of the junctions conduct noticeably at 0.6 V. Current starts to rise at 0.7 V. At 0.8 V, one junction draws almost 80 mA. At 1 V, the dc resistance is on the order of 2 or 3 Ω, and the transistor draws almost 1 A. In a germanium transistor, noticeable current flow occurs at about 0.3 V.

From a troubleshooting standpoint, bias measurements provide a clue to the performance of solid-state oscillators. However, bias measurements do not provide proof positive. The one sure test of an oscillator is measurement of the wave form on an oscilloscope. If the wave form is present and is of the right shape, amplitude, and time duration (frequency), the oscillator is operating properly.

Usually, the collector is the best place to measure a solid-state oscillator wave form. Keep in mind that the scope must have a *band pass greater*

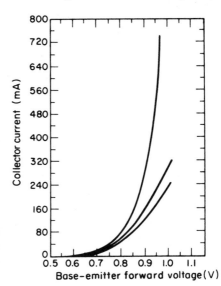

FIGURE 3-47 Characteristic curves for silicon transistor junctions (collector current flow versus base-emitter forward voltage)

than the oscillator frequency. If not, the wave form will be distorted or will not pass at all. If it is necessary to measure a frequency higher than the scope band pass, use an RF probe to test the oscillator.

3-4.2 Basic RF Tests

The following paragraphs describe test procedures for RF amplifiers used in transmitters. The first paragraphs are devoted to test and measurement procedures for the resonant circuits used at radio frequencies (resonant-frequency measurements, Q measurements, and so forth). The remaining sections cover test procedures for complete RF amplifiers used in transmitters.

For best results, RF amplifiers are tested with all components soldered in place, both before and after troubleshooting. This will show whether there is any change in circuit characteristics resulting from the physical relocation of components during troubleshooting or repair procedures.

Often, there is capacitance between components, from components to wiring, and between wires. These stray components can add to the reactance and impedance of circuit components. When the physical locations of parts and wiring are changed, the stray reactances change and alter circuit performance.

3-4.3 Basic RF Voltage Measurement

As we discussed in Chapter 2, when the voltages to be measured are at radio frequencies and are beyond the frequency capabilities of a meter or scope, an RF probe is required. Such probes rectify the RF signals into a dc output that is almost equal to the peak RF voltage. The dc output of the probe is then applied to the meter or scope and is displayed as a voltage readout in the normal manner.

If a probe is available as an accessory for a particular meter or scope, use that probe rather than any homemade probe. The manufacturer's probe is matched to the meter or scope in calibration, frequency compensation, and the like. If a probe is not available for a particular meter, the following notes can be used to make a probe for RF voltage measurement.

The half-wave probe, shown in Fig. 3-48, provides an output to the meter or scope that is approximately equal to the peak value of the voltage being measured. Because most meters are calibrated to read in rms values, the probe output must be reduced to 0.707 of the peak value by means of R_1. A variable (noninductive) resistor can be substituted for R_1 during calibration and then replaced by a fixed resistor of the correct value.

To calibrate the probe, apply an RF voltage of precise, known amplitude to the RF input terminals. Adjust R_1 until the meter reads 0.707 times the known input voltage. For example, with 10 V RF at the input, adjust R_1

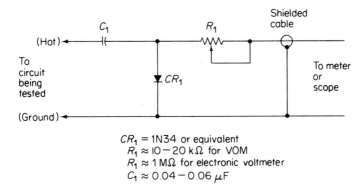

CR_1 = 1N34 or equivalent
$R_1 \approx 10-20$ kΩ for VOM
$R_1 \approx 1$ MΩ for electronic voltmeter
$C_1 \approx 0.04-0.06$ μF

FIGURE 3-48   Half-wave probe for RF measurements

for a reading of 7.07 V on the meter. Replace variable R_1 with a fixed resistance of corresponding value (or leave R_1 set at the correct value). Repeat the test over the anticipated frequency range.

The probe shown in Fig. 3-48 should provide satisfactory results up to about 250 MHz. Beyond that frequency, always use the probe supplied with the meter or scope. Keep in mind that the meter must be set to read direct current because the probe output is direct current. (If the RF input signal is modulated, the probe output may be pulsating direct current.)

3-4.4 Measuring LC Circuit Resonant Frequency

The circuit for measuring resonant frequency of an LC circuit is shown in Fig. 3-49.

To use the circuit, adjust the unmodulated RF generator's output amplitude for a convenient indication on the meter. Then, starting at a frequency well below the lowest possible frequency of the LC circuit, slowly increase the generator-output frequency.

For a parallel-resonant LC circuit, watch the meter for a maximum or peak indication.

For a series-resonant LC circuit, watch the meter for a minimum or dip indication.

The resonant frequency of the LC circuit is the one at which there is a maximum (for parallel) or minimum (for series) indication on the meter.

Note that there may be peak or dip indications at harmonics of the resonant frequency. The test is most efficient when the approximate resonant frequency is known.

To broaden the response (so that the peak or dip can be approached more slowly), increase the value of R_L from 100 kΩ. (An increase in R_L lowers the LC circuit Q). To sharpen the response, lower the value of R_L.

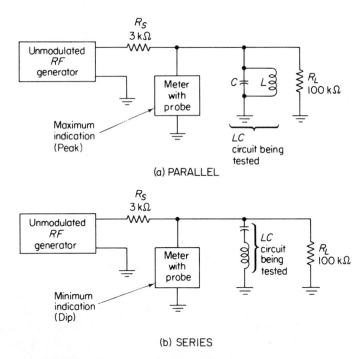

(a) PARALLEL

(b) SERIES

FIGURE 3-49   Measuring resonant frequency of LC circuits

3-4.5 Measuring Inductance of a Coil

The circuit for measuring inductance of a coil is shown in Fig. 3-50. To use the circuit, adjust the unmodulated RF generator's output amplitude for a convenient indication on the meter. Then, starting at a frequency well below the lowest possible resonant frequency of the LC combination being tested, slowly increase the generator-output frequency.

The resonant frequency of the LC circuit is the one at which there is a *maximum* indication on the meter. Using this resonant frequency and a known capacitance value, calculate the unknown inductance using the equations given in Fig. 3-50. To simplify calculations, use a convenient capacitance value, such as 100 picofarad (pF) or 1,000 pF.

Note that when a known inductance value is available, the procedure can be reversed to find an unknown capacitance value.

Increase the value of R_L to broaden the peak indication if desired or decrease R_L to sharpen the peak.

3-4.6 Measuring Self-Resonance and Distributed Capacitance of a Coil

There is distributed capacitance in any coil, which can combine with the coil's inductance to form a resonant circuit. Although the self-resonant frequency may be high in relation to the operating frequency at which the

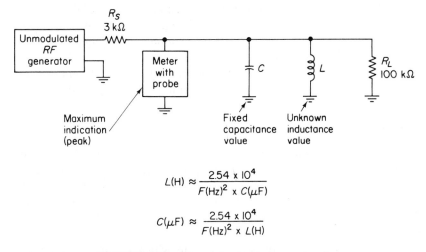

$$L(H) \approx \frac{2.54 \times 10^4}{F(Hz)^2 \times C(\mu F)}$$

$$C(\mu F) \approx \frac{2.54 \times 10^4}{F(Hz)^2 \times L(H)}$$

FIGURE 3-50   Measuring inductance of coil

coil is used, it may be near a harmonic of that operating frequency. This limits the usefulness of the coil in an LC circuit. Some coils, particularly RF chokes used in transmitters, may have more than one self-resonant frequency.

The circuit for measuring self-resonance and distributed capacitance of a coil is shown in Fig. 3-51.

To use the circuit, adjust the unmodulated RF generator's output amplitude for a convenient indication on the meter. Tune the generator over its entire frequency range, starting at the lowest frequency. Watch for either peak or dip indications on the meter. Either a peak or dip indicates that the inductance is at a self-resonant point. The generator-output frequency at this point is the self-resonant frequency (or a harmonic).

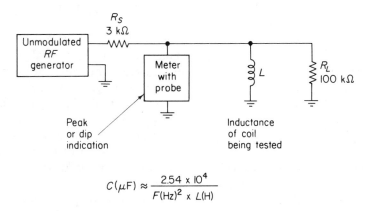

$$C(\mu F) \approx \frac{2.54 \times 10^4}{F(Hz)^2 \times L(H)}$$

FIGURE 3-51   Measuring self-resonance and distributed capacitance of coil

Make certain that peak or dip indications are not the result of changes in generator-output level. Cover the entire frequency range of the generator or at least from the lowest up to the third harmonic of the highest frequency involved in circuit design or operation.

Once the resonant frequency (or frequencies) has been found, calculate the distributed capacitance using the equation given in Fig. 3-51.

3-4.7 Measuring Q of Resonant Circuits

The Q of a resonant circuit sets the circuit band widths. That is, a high Q circuit has a narrow band width (sharp tuning); whereas a low Q circuit has a wide band width (broad tuning). The most practical measurement of resonant circuit Q is to measure the band width at the resonant frequency. The circuits for measuring band width (or Q) of resonant circuits are shown in Fig. 3-52.

Figure 3-52(a) shows the test circuit in which the signal generator is connected directly to the input of a complete stage; Fig. 3-52(b) shows the indirect method of connecting the signal generator to the input.

When the stage or circuit has sufficient gain to provide a good reading on the meter with a nominal output from the generator, the indirect method (with isolating resistor) is preferred. Any signal generator has some output impedance (typically 50 Ω). When this resistance is connected directly to the tuned circuit, the Q is lowered, and the response becomes broader. (In some cases, the generator-output impedance can seriously detune the circuit.)

Figure 3-52(c) shows the test circuit for a single component (such as an RF transformer).

When the resonant circuit is normally used with a load, the most realistic Q measurement is made with the circuit terminated in that load value. A fixed resistance can be used to simulate the load. The Q of a resonant circuit is often dependent upon the load value.

To use the circuit, adjust the unmodulated RF generator's output amplitude for a convenient indication on the meter. Tune the signal generator to the approximate resonant frequency of the circuit; then, tune the generator for maximum or peak reading on the meter. Note the generator frequency.

Tune the generator below resonance until the meter reading is 0.707 of the maximum reading. Note the generator frequency (frequency F_2). To make the calculation more convenient, adjust the generator-output level so that the meter reading is some even value (such as 1 V or 10 V) after the generator is tuned for maximum. This will make it easy to find the 0.707 mark.

Tune the generator above resonance until the meter reading is 0.707 of the maximum reading. Note the generator frequency (frequency F_1).

Calculate the circuit Q using the equation given in Fig. 3-52.

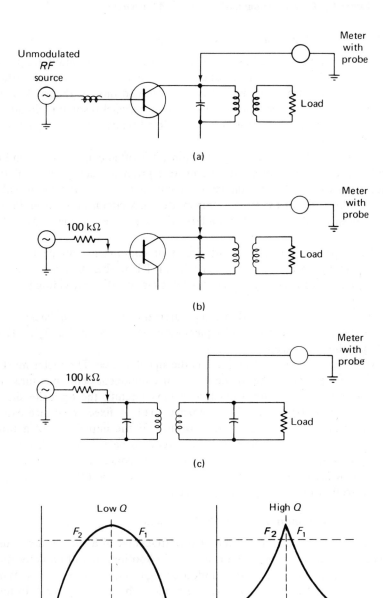

$$Q = \frac{F_R}{F_1 - F_2}$$

F_R = peak resonant frequency

FIGURE 3-52 Measuring Q of resonant circuits

3-4.8 Measuring Impedance of Resonant Circuits

Any resonant circuit has some impedance at the resonant frequency. The impedance changes with frequency. This includes transformers (tuned and untuned, RF transmitter tank circuits, and so on). In theory, a series-resonant circuit has zero impedance, and a parallel-resonant circuit has infinite impedance. In practice, it is impossible because there is always some resistance in the circuit.

In some RF troubleshooting situations, it is often convenient to find the actual impedance of a resonant circuit at a given frequency. Also, it may be necessary to find the impedance of a component so that the circuit values can be confirmed. For example, a transmitter tank circuit presents an impedance at both its primary and secondary windings. These values may not be specified.

The impedance of a resonant circuit or component can be measured using a signal generator and a meter with an RF probe. An electronic voltmeter provides the least loading effect on the circuit, thus providing the most accurate indication.

The procedure for impedance measurement at radio frequencies is the same as the procedure at audio frequencies, as discussed in Sec. 3-2.8. However, the following exceptions apply:

An RF generator must be used as the signal source. The meter must be provided with an RF probe. If the circuit or component being measured has both an input and an output (such as a tank circuit), the opposite side or winding must be terminated in its normal load. A fixed resistance can be used to simulate the normal load resistance. If the impedance of a tuned circuit is to be measured, first tune the circuit to peak (for parallel circuit) or dip (for series circuit); then, measure the impedance at resonance. Once the resonant impedance is found, the generator can be tuned to other frequencies to find the corresponding impedance.

3-4.9 Testing Transmitter RF Amplifier Circuits

During normal operation, the final output stage of a transmitter is usually connected to an antenna. During troubleshooting, it is often convenient to disconnect the antenna so that undesired signals are not broadcast. When the antenna is disconnected, it must be replaced by a *dummy load*. In most troubleshooting situations, a simple fixed resistor can serve as the dummy load. Do not use a wirewound resistor. All wirewound resistors have some inductance (in addition to the resistance), and the inductance changes with the frequency. A dummy load should present pure resistance to the transmitter's final output. Figure 3-53 shows typical connections for a dummy-load resistance. The load resistance must be equal to the antenna impedance (typically about 50 Ω for communications transmitters and antennas).

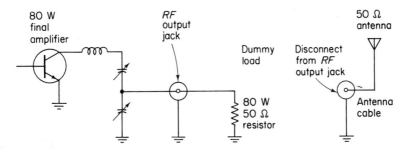

FIGURE 3-53   Connections for dummy load used during trouble-shooting of transmitter RF circuits

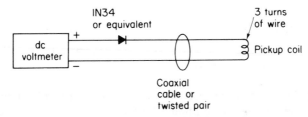

FIGURE 3-54   Circuit for pickup and measurement of RF signals (used for troubleshooting transmitter circuits)

It is possible to test and adjust transmitter RF amplifiers using a meter with an RF probe. If an RF probe is not available, or as an alternative, it is possible to use a test circuit such as the one shown in Fig. 3-54. This circuit is essentially a pickup coil (which is placed near the RF amplifier inductance) and a rectifier that converts the radio frequency into a dc voltage for measurement on a meter.

Figure 3-55 shows the basic circuit for test and measurement of RF amplifier circuits. (The circuit also provides for test of the oscillator.) If the amplifier being measured does not have an oscillator, a drive signal must be supplied by means of a signal generator. Use an unmodulated signal at the correct frequency.

Connect the pickup circuit to each amplifier stage in turn. Start with the first stage (this will be the oscillator if the circuit being tested is a complete transmitter), and work toward the final or output amplifier stage.

A voltage indication should be obtained at each stage. Usually, the voltage indication increases with each amplifier stage. Some stages may be frequency multipliers and therefore provide no voltage amplification.

If a particular amplifier stage is to be tuned, adjust the tuning control for a maximum reading on the meter. If the stage is to be operated with a load (such as the final amplifier into an antenna), the load should be connected,

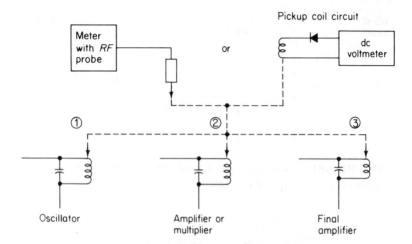

FIGURE 3-55   Connections for test of RF amplifier circuits during transmitter troubleshooting

or a simulated load should be used. A fixed resistance provides a good simulated load at frequencies up to about 250 MHz.

It should be noted that this tuning method or measurement technique does not guarantee that each stage is at the desired operating frequency. The method does show that a signal is present and that the circuit is tuned for peak. However, it is possible to get maximum readings on harmonics. Fortunately, it is conventional to design RF transmitter amplifier circuits so that they will not tune to both the desired operating frequency and a harmonic. Generally, RF amplifier tank circuits tune on either side of the desired frequency but not to a harmonic (unless the circuit is seriously detuned or the design calculations are hopelessly inaccurate).

3-4.10 Modulation Measurement

An oscilloscope can be used to display the carrier of an AM wave at the output of a transmitter. There are two basic methods: direct measurement of the modulation envelope and conversion of the envelope to a trapezoidal pattern. The trapezoidal method is the most effective because it is easier to measure straight-line dimensions than it is to measure curving dimensions. Also, nonlinearity in modulation can be checked easily with the straight-line trapezoid. With either method, the percentage of modulation can be calculated from the dimensions of the modulating pattern.

Direct measurement of the modulation envelope. The equipment is connected for direct measurement of the modulation envelope, as shown in Fig. 3-56. If the vertical channel response of the scope is capable of handling the transmitter's output frequency, the output is applied through the scope's

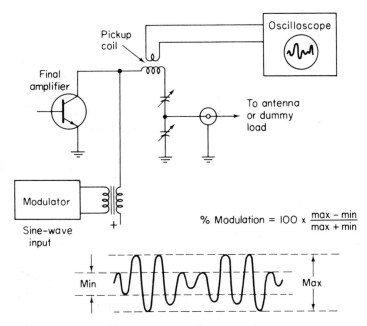

FIGURE 3-56   Connections for direct measurement of modulation envelope

vertical amplifier. If it is not, the transmitter output must be applied directly to the vertical deflection plates of the scope's cathode-ray tube.

When the transmitter is amplitude-modulated with a sine wave, the resultant wave form is similar to that shown in Fig. 3-56. If the transmitter is operating properly, the modulation envelope should reflect the sine-wave modulation signal. By measuring the vertical dimensions MAX and MIN, shown in Fig. 3-56, the percentage of modulation can be calculated.

Trapezoidal measurement of the modulation envelope. The equipment is connected for trapezoidal measurement of the modulation envelope, as shown in Fig. 3-57. With this method, the scope's amplifiers are not used. Instead, both the horizontal and vertical connections are made directly to the scope's cathode-ray tube. When the transmitter is amplitude-modulated with a sine wave, the resultant wave form patterns are similar to those shown in Fig. 3-57.

The scope's display width is adjusted by means of resistor R_1. The height of the scope's display is adjusted by varying the coupling between the pickup coil and the output tank circuit of the transmitter.

Figures 3-57(c), (d) and (e) show typical wave form patterns for 50 percent modulation, 90 to 95 percent modulation, and overmodulation (i.e., over 100 percent modulation), respectively. By measuring the vertical dimen-

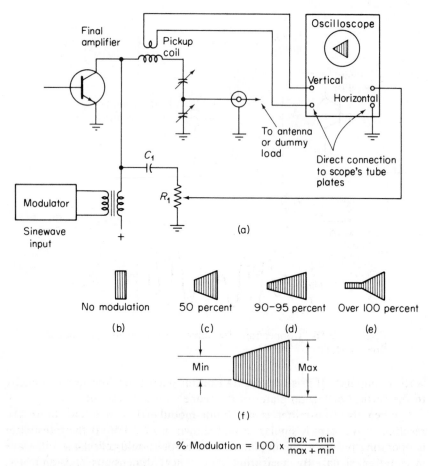

FIGURE 3-57   Connections for trapezoidal measurement of modulation envelope

sions MAX and MIN, shown in Fig. 3-57(f), the percentage of modulation can be calculated.

3-4.11 Example of Transmitter Troubleshooting

This step-by-step troubleshooting example involves locating the defective component in a solid-state radio transmitter and then repairing the trouble. The transmitter is considered a single-unit, multicircuit piece of equipment consisting of two circuit groups. Although two circuit groups are involved, you can skip the *localize* step because there is only one unit.

General instructions. The servicing block diagram for the transmitter is shown in Fig. 3-58. Test points *A* to *L* on the block diagram are more accurately defined on the schematic diagram, as shown in Fig. 3-59. Note that

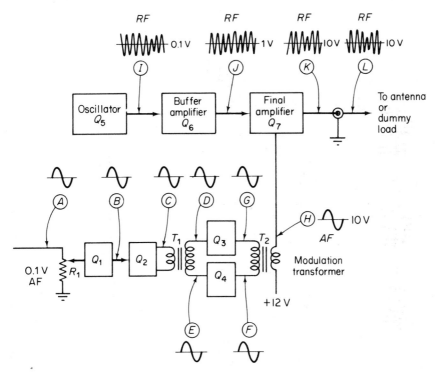

FIGURE 3-58   Servicing block diagram of transmitter

the modulator portion of the transmitter is similar to that described for the audio-amplifier troubleshooting example (Sec. 3-2.17), with one major exception. The output of the modulator is used to modulate the RF section (instead of driving a loudspeaker). A modulation transformer is used instead of an output transformer. Power for the final amplifier of the RF section is applied through the secondary windings of the modulation transformer. Thus, any audio signal present at test point A is amplified by the modulator and serves to amplitude-modulate the RF output.

The servicing block diagram and the schematic diagram both show test points. Thus, it is not necessary for you to determine the location of the test points before troubleshooting. However, in practical situations, there is nothing that says you cannot add test points of your own. For example, the diagrams show that there should be an RF signal of approximately 1 V at test point J (the collector of amplifier Q_6). The same signal should also appear at the base of final amplifier Q_7, even though no test point is indicated. If, during troubleshooting, you find that there is a good signal at J but none at the base of Q_7, you have traced a fault to that portion of the circuit (probably to the tuning network or possibly to the RF choke or Q_7). In any event, do *not* feel limited to the test points specified on service literature.

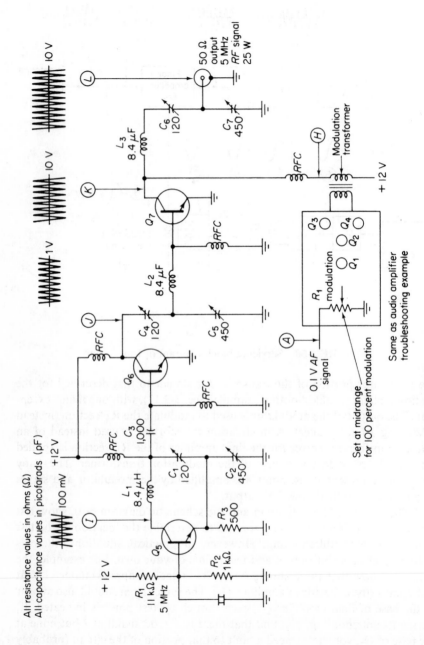

FIGURE 3-59 Schematic digaram of transmitter

No voltage or resistance information is available except for the $+12$ V supply voltage found on the schematic. However, you should be able to calculate the voltages found at the transistor elements. The collectors of the three RF section transistors (Q_5, Q_6, and Q_7) are all connected to the $+12$ V through RF chokes. Such chokes generally have very little dc resistance and thus produce very little voltage drop (typically a fraction of a volt drop). Thus, it is reasonable to assume that the dc voltage at the collectors is about $+12$ V (or slightly less).

The emitters of Q_6 and Q_7 are connected directly to ground (which is typical for npn RF amplifiers). The emitter's dc voltage (and resistance) should be 0 V. The same is probably true of the Q_6 and Q_7 bases. The only resistance from the bases should be a few ohms produced by the RF chokes. It is possible that some dc voltage might be developed across the chokes, but it is not likely.

The base of Q_5 has a fixed dc voltage applied through the voltage-divider network of R_1 and R_2. The ratio of R_1 to R_2 indicates that the voltage drop across R_2 is about 1 V; thus, the base of Q_5 is about 1 V. The emitter of Q_5 should be about 0.5 V less than the base, or about $+0.5$ V with respect to ground. The resistance to ground from the emitter of Q_5 should be equal to R_3, or about 500 Ω.

The schematic shows that the output is 25 W into a 50 Ω antenna. If a dummy load is used (as it should be during routine troubleshooting), use a 25 W (or larger) resistor with a value of 50 Ω. Do not use a wirewound resistor. (It is possible to use a 25 W lamp as the dummy load, but the fixed resistance is generally preferred.)

Both the block diagram and the schematic show that the output frequency is 5 MHz. The schematic shows that the crystal is also 5 MHz. This means that the three RF stages are tuned to the same frequency. In some transmitters, one (or more) of the RF amplifier stages will be tuned to a harmonic of the fundamental frequency. Thus, the stages act as frequency multipliers.

Generally, this is of concern only when you are trying to measure the frequency at each stage. From a practical standpoint, it is usually possible to troubleshoot a transmitter without actually knowing the frequencies at each stage. You are more concerned that radio frequency is present and is of the correct amplitude. Even when making final tuning adjustments, you tune for a peak indication at each stage, rather than to a specific frequency.

Capacitors C_1, C_2, and C_4 to C_7 are tuning adjustments. Keep in mind that these are *adjustment* controls, rather than operating controls. That is, these capacitors are accessible for adjustment only after covers have been removed and are not touched during routine operation. The only operating control is R_1, the modulation potentiometer. The schematic indicates that the RF output will be modulated 100 percent when R_1 is set to the approximate midrange and there is a 100 mV audio signal at the modulator input (test point *A*).

With this wealth of information to draw upon, you are ready to plunge into the troubleshooting by determining the symptoms.

Determine the symptoms. In some troubleshooting work, you will find that the equipment was never properly tuned, adjusted, aligned, or otherwise put into operation properly. Some troubleshooters approach each new problem with that assumption. However, in the majority of troubleshooting problems, you will find that the equipment was working satisfactorily for a time before the trouble occurred. This is the case in this example. The transmitter had been working properly for many months, but the operator reports to you that there is now no output. His transmission cannot be heard by any stations tuned to the operating frequency (5 MHz). What is your first step?

You could start by checking each circuit to determine which one is not operating. This rates a definite *no.* You will *eventually* locate the faulty circuit with this procedure. However, it is not a logical approach and would probably require several unnecessary tests before the trouble was isolated to a circuit.

You could perform the tuning procedure to determine which circuit group does not perform properly. This would be even worse than checking each stage. The procedure is not a logical step in troubleshooting. Although you *might* locate the trouble area with this procedure, it would probably require many unnecessary steps. Each step should provide the most information with the least amount of testing.

You should check the output. Notice that we did not tell you *how* to check the output at this time. You have two basic choices for checking the output: with test equipment and without test equipment. The output can be checked by tuning a receiver to the operating frequency, turning on the transmitter, and trying to transmit a signal (probably a voice into a microphone at the modulator input). This is a quick and easy test, but it does not prove much.

Even if you hear the voice transmission on the receiver, it is possible that the transmitter output is low. A weak transmission might be picked up by a receiver nearby but could not be heard over the normal communications range. If you do not hear the transmission on the receiver, it only proves that the operator was right, but it does not give you any clue to which section of the transmitter is at fault.

A more practical method of determining the symptoms is to monitor the output of the final stage Q_7. Note that the transmitter is not provided with any built-in operating indicators. That is, there is no RF output indicator (a lamp or meter) and no modulation meter. Thus, you must use external test equipment. Before making any such tests, disconnect the antenna, and replace it with a dummy load.

You can monitor the RF output at test points K or L with a meter or scope, using an RF probe. However, this will only prove the presence (or absence) of RF signals. The most satisfactory test is to use a scope connected as shown in Fig. 3-60. (The scope's vertical amplifier must be capable of passing the 5 MHz RF signals.)

For a complete output test, the transmitter should be modulated with an audio tone. You could use voice, but the results will be inconclusive. A better test is to set R_1 to midrange and apply a 100 mV audio tone (say, at 1 kHz). As shown on the schematic, this should result in 100 percent modulation (i.e., the modulation envelope, Fig. 3-60, should drop to zero between peaks). The percentage of modulation should be controlled by R_1. Although there are no specifications given on the schematic, it is reasonable to assume

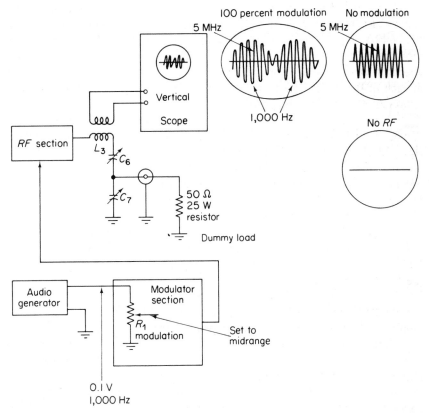

FIGURE 3-60 Connections for testing transmitter output

that with R_1 set to full ON, the percentage of modulation will be in excess of 100. This should be verified during the output check.

Now, assume that there is no RF indication on the scope (no vertical deflection whatsoever, with R_1 set at any position). In which group do you think the trouble is located?

Modulator section. If you choose the modulator section, you will probably not succeed as a troubleshooter, or you are not paying attention, or you simply do not understand transmitters. The vertical deflection on the scope is produced by the RF signal; the shape of the signal is determined by the modulation. Thus, with no vertical deflection, there is no radio frequency, and trouble is traced to the RF section. The setting of R_1 has no effect on the presence or absence of radio frequency.

RF section. You have made a logical choice. The next step is to isolate the trouble to one of the circuits (oscillator Q_5, buffer Q_6, or power amplifier Q_7) in the defective circuit group. What is your next most logical test point?

You could check the supply voltage for the RF circuits. This approach has some merit. Note that Q_5 and Q_6 receive their collector voltage directly from the $+12$ V supply line; whereas Q_7 gets its collector voltage through the modulation transformer's secondary winding. Thus, to make a complete check of supply voltage for the RF circuits, you must measure the voltage at each of the three collectors. Of course, if the voltage at any one collector is present and correct, the power supply is good.

You could check for radio frequency at test point I. This is not a bad choice. If there is no RF signal at I, the trouble is traced to the oscillator (Q_5 and associated parts). This would be a lucky guess. If there is an RF signal at I, you know that the oscillator is operating, but you eliminate only one circuit as a possible trouble area. Therefore, it is not the most logical choice.

You should check for radio frequency at test point J. This is the most logical choice and can be done with a meter and RF probe, as shown in Fig. 3-61. As an alternate, a scope and pickup coil can be used, as shown in Fig. 3-61(b). Of course, the scope's vertical amplifier must be capable of passing the 5 MHz RF signal.

If there is a good RF signal indication at J, both Q_5 and Q_6 can be considered to be operating properly, and the trouble is traced to the power amplifier Q_7. You also eliminate the power supply because it must be good if Q_5 and Q_6 are functioning normally. However, you do not eliminate the lead between the power supply and the collector of Q_7 (which is a separate path from that between supply and the Q_5 and Q_6 collectors).

If there is no RF indication at J, the trouble is traced to Q_5 or Q_6. The next step is to check for radio frequency at I.

Now, assume that there is a good RF indication at J. The power ampli-

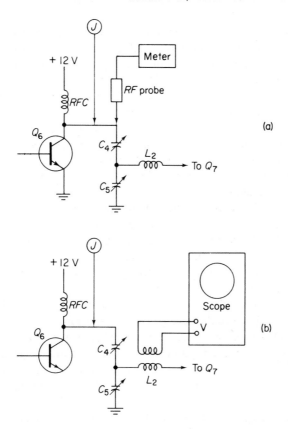

FIGURE 3-61   Checking for RF signal at test point *J*

fier (Q_7 and associated circuit parts) can be bracketed as the faulty circuit because the input signal is good and the output is bad.

Locate the specific trouble. Now that the trouble is isolated to the circuit, it must be located. The first thing you do is perform a visual inspection of the power amplifier. Assume that there is no *apparent* sign of where the trouble could be located. There is no sign of overheating, and all components (as well as wiring) appear proper. What is your next step?

You could make an in-circuit test of Q_7. This would be difficult because Q_7 is operated class *C* and will not respond to the usual in-circuit forward-bias tests (as discussed in Sec. 3-1.4). There is no forward bias applied to Q_7, so you cannot remove the bias. If you attempt to apply bias, the voltage relationship between emitter and collector will probably not change because there is no dc load. An in-circuit transistor tester might prove that the transistor is good at audio frequencies. However, in-circuit testers are generally useless at radio frequencies.

You could make a substitution test of Q_7. This would be more satisfactory than any in-circuit test. And it is possible that you may have to substitute Q_7 before you have located the fault. However, there are more convenient tests to be made at this time.

You could check the resistance at all elements of Q_7. Although it will probably be necessary to check resistance and continuity before you are through, resistance checks at this time will prove little. The resistance-to-ground at the emitter of Q_7 should be zero. The base resistance-to-ground should also be near zero. (The RF choke winding may show a few ohms on the lowest ohmmeter scale.) Only a high resistance-to-ground at the base and emitter would be significant.

You do not know the correct resistance-to-ground for the collector of Q_7. Of course, you could guess. The resistance of the Q_7 collector should be substantially the same as that for the Q_5 and Q_6 collectors, plus any resistance in the modulation transformer's secondary winding. Thus, if the Q_7 resistance is slightly higher than the Q_5 and Q_6 collector resistance, you can *assume* that the value is correct.

You should check the voltage at all elements of Q_7 first. The voltage at the base and emitter should be zero (we are speaking of dc voltage, not RF signal voltage). The dc voltage at the Q_7 collector should be about $+12$ V. If the voltages are all good, you can skip the resistance-to-ground measurements. However, you still may have to make continuity checks if the voltages are abnormal. Let us examine possible faults that are indicated by abnormal voltages.

Large dc voltages at base or emitter. If there are large dc voltages at the base or emitter of Q_7, this indicates that the elements are not making proper contact with ground. For example, if there was a high-resistance solder joint between the Q_7 emitter and ground, it is possible for a dc voltage to appear at the emitter. Or if the emitter-ground connection were completely broken, the emitter would be floating and show a dc voltage.

No dc voltage at collector. If the collector shows no dc voltage, the fault is probably in the RF choke or the modulation transformer's winding. This requires a continuity check. Check the dc voltage at test point H. If the voltage is correct at H but not at K (the collector of Q_7), the RF choke is at fault. If the voltage is absent at H, the fault is in the transformer winding.

Now, to summarize this troubleshooting example, let us assume that the trouble is caused by an open L_3 coil winding. This will not affect dc voltage or resistance. Substitution of Q_7 will not cure the problem. These are the kinds of problems you find in *real* troubleshooting: Everything *appears* to be good, but the equipment will not work!

To solve such problems, you must make point-to-point *continuity checks*. In this case, if you checked from point K to the top of C_6, you would find the open coil winding.

Index

235

V

Vacuum tube testing, 66
Verifying trouble, 38
Visual inspection, 60
Voltage:
  divider probe, 139
  effects of, 169
  gain (amplifier), 174
  measurements, 8, 66, 105, 151
  measurements (IC), 164
  test, 63
Voltmeter, 85

Voltmeter, electronic, 92
VOM (voltohmmeter), 82
VTVM (vacuum tube voltmeter), 82

W

Waveforms, 8
  measurements, 64, 127
  test, 63
Wheatstone bridge, 111
Wiring diagram, 6